RAND NATIONAL DEFENSE RESEARCH INSTITUTE

A Policy Analysis of Reserve Retirement Reform

Beth J. Asch, James Hosek, Michael G. Mattock

Prepared for the Office of the Secretary of Defense

The research described in this report was prepared for the Office of the Secretary of Defense (OSD). The research was conducted within the RAND National Defense Research Institute, a federally funded research and development center sponsored by OSD, the Joint Staff, the Unified Combatant Commands, the Navy, the Marine Corps, the defense agencies, and the defense Intelligence Community under Contract W74V8H-06-C-0002.

Library of Congress Cataloging-in-Publication Data is available for this publication.

ISBN: 978-0-8330-7809-4

Published 2013 by the RAND Corporation
1776 Main Street, P.O. Box 2138, Santa Monica, CA 90407-2138
1200 South Hayes Street, Arlington, VA 22202-5050
4570 Fifth Avenue, Suite 600, Pittsburgh, PA 15213-2665
RAND URL: http://www.rand.org/
To order RAND documents or to obtain additional information, contact
Distribution Services: Telephone: (310) 451-7002;
Fax: (310) 451-6915; Email: order@rand.org

Preface

As the burden of defense borne by reserve forces has increased, more attention has been paid to differences between the compensation systems for the reserve and active components. One particular emphasis is on the retirement systems, a key difference being that reserve members who complete 20 years must wait until age 60 to draw their retirement benefits whereas active members can draw their benefits immediately.

This monograph provides a policy analysis of reserve retirement reform. It compares the reserve and active retirement systems, discusses the importance of structuring compensation to enable flexibility in managing active and reserve manpower, describes how the debate over reserve retirement reform has differed from active component retirement reform debate, and considers obstacles to reform and how they might be overcome. It also provides a quantitative assessment of several past congressional proposals to change the reserve retirement system in terms of their effects on reserve participation and personnel costs. None of the proposals were actually legislated, although discussions surrounding them led to the fiscal 2008 National Defense Authorization Act, which changed reserve retirement eligibility so that National Guard and Reserve members can begin to receive retirement benefits three months sooner than age 60 for each cumulative period of 90 days served on active duty. These proposals represent the types of changes that continue to be advocated by those who desire to align the reserve retirement system more closely with the active retirement system. The analysis in the report shows the effects of these proposals on reserve participation, as well as their possible effects on active retention, the transition from active to reserve forces, and cost

Early results of this analysis were presented to the Office of the Secretary of Defense (OSD) and helped to inform the policy discussion about reserve component (RC) retirement changes. Early results were also presented at a conference on managing reserve forces under a continuum of service concept.

The present volume takes advantage of recent developments in our modeling capability that serve to increase the precision and consistency of the parameter estimates and thereby produce more-reliable policy simulations. The findings reported here come from these new estimates and policy simulations. The improved capability emerged as we conducted research for the 11th Quadrennial Review of Military Compensation, and the updated results are consistent with the early results presented to OSD.

The policy analysis in this monograph may be of interest to defense manpower policymakers concerned with the adequacy of reserve pay and the form it should take. The model and estimation techniques may interest a technical audience.

This research was conducted within the Forces and Resources Policy Center of the RAND National Defense Research Institute, a federally funded research and development center sponsored by the Office of the Secretary of Defense, the Joint Staff, the Unified Combatant Commands, the Navy, the Marine Corps, the defense agencies, and the defense Intelligence Community.

For more information on the RAND Forces and Resources Policy Center, see http://www.rand.org/nsrd/ndri/centers/frp.html or contact the director (contact information provided on the web page).

Contents

Figures

Tables

Summary

The greater usage of the reserve components (RC) in an operational capacity to support missions such as those in Iraq and Afghanistan has led members and policymakers to pay more attention to the compensation and personnel systems that support RC management and how those systems compare to those supporting the active components (AC). The research presented in this report focuses on one aspect of RC compensation, the retirement system, which differs from the AC's. While both reward members who complete 20 years of service (YOS) with a retirement annuity based on years of service and basic pay, some say the current system is inequitable because a reservist who completes 20 creditable years of service may not draw retirement benefits immediately after retiring from the military, as an AC retiree can.

The report considers the myriad issues surrounding the debate over reforming the RC system and provides quantitative assessments of specific proposals to change the system. The proposals we evaluate were introduced by Congress between 2003–2006 but never legislated, and the idea beyond them, namely increasing the alignment of the RC and AC retirement systems, is a topic of ongoing debate. Thus, our analysis is intended to inform current as well as future debates about changing the RC retirement system.

The specific proposals we consider are bills introduced in the 108th and 109th Congress that aimed to decrease the age at which reservists may draw retirement benefits:

- H.R. 331 (108th Congress) provided benefits immediately upon retiring from the military with 20 or more YOS, regardless of one's age ("immediate annuity" proposal).
- S. 32, S. 38/H.R. 1169, H.R. 783 (109th Congress) decreased the age of first receipt from age 60 to age 55 ("age 55").
- S. 337/H.R. 558 (109th Congress) decreased the age of first receipt by one year for every two years served beyond 20 years: A reservist may retire at age 60 with 20 YOS, at age 59 with 22 YOS, and so forth to age 53 with 34 YOS ("sliding-scale").

The first proposal—the immediate annuity proposal—would perfectly align the RC retirement system with the AC system to the extent that under this proposal both would provide retirement benefits to vested personnel as soon as they left service. This is an idea that is often advocated by those who would like to see identical retirement systems. The other two proposals are more modest, although both move toward reducing the difference between when members can first draw benefits. The age-55 proposal would slip the retirement age from 60 to 55, while the sliding-scale proposal would reduce the age depending on years of service.

We employ a dynamic programming model of AC retention and RC participation, known as the dynamic retention model (DRM). We estimate the DRM using Defense Manpower Data Center Work Experience File (WEX) data that tracks through 2010 the individual careers of AC members—including for those who join, their time in the RC—for those who entered active service in either 1990 or 1991. We use the model estimates to simulate the effects of the proposals in a steady state. The DRM not only shows quantitative effects of the proposals on RC participation, but also the effects that might occur on AC retention.

In addition to the quantitative analysis, we consider the broader issues surrounding reserve retirement reform, including consideration of AC/RC retirement benefit equity as well as other goals of reserve retirement reform, such personnel management flexibility. We also discuss reserve retirement reform in the broader context of AC retirement reform, and we identify approaches to implementing military retirement reform that might overcome the obstacles that have hampered past efforts.

Key Findings on the Congressional Proposals

We briefly summarize the main findings of our policy simulations. First, we consider the effects on reserve participation. The age-55 and sliding-scale proposals would have a small effect on the percentage of active members who later join the reserves (1–3 percent for enlisted and 1–6 percent for officers), while the immediate annuity proposal would have a more substantial effect (5–20 percent for enlisted and 15–35 percent for officers). This is not surprising because the immediate annuity proposal is more generous than the other proposals and so would have larger effects. Similarly, the age-55 and sliding-scale proposals would only cause small changes in reserve retention while the immediate annuity proposal would cause large effects, especially in the experience mix of RC members. Under the immediate annuity proposal, mid-career reserve participation would increase, before the vesting point of 20 YOS, but subsequently decline after 20 YOS. This pattern reflects a behavioral response to the opportunity to receive reserve retirement benefits immediately rather than waiting until age 60. All three proposals would increase the percentage of RC members qualifying for RC retirement benefits, typically by 2–10 percent for the age-55 and sliding-scale proposals and by

40–80 percent for enlisted and 30–45 percent for officers for the immediate annuity proposal.

We also estimated the cost effects of the proposals. We find that the age-55 and sliding-scale proposals would increase cost per member by about 4 percent. The immdiate annuity would generate a substantial increase in cost per member, nearly 25 percent. These increases would occur because retirement costs would increase as more RC members qualify for RC retirement benefits under these proposals and because all three proposals offer more-generous benefits than the current system.

On net, we find that the proposals are not cost-effective methods of sustaining the overall size of the RC prior-service force. Our finding that the cost per member increases under each proposal indicates that these proposals are more-expensive ways of maintaining the same RC force size.

Equity, Deployment, and Force-Management Flexibility

In addition to our quantitative assessment of the three specific proposals, we took a broader perspective on the issues surrounding reserve retirement reform and changing the age at which reservists can begin receiving benefits. Three of those issues are inequity of reserve pay and benefits with respect to active pay and benefits, recognition of more-frequent and longer deployments for reservists, and the role of the retirement system in facilitating the flexible management of personnel.

The main apparent inequity of the RC retirement system is that reserve retirement benefits begin at age 60 whereas active benefits begin immediately upon separation with 20 or more YOS. However, equity comparisons should extend beyond when retirement benefits start, because active and reserve service differs in important ways. AC personnel serve full time, are always on call for duty, and are likely to be deployed more frequently. They cannot hold a full-time civilian job and therefore cannot accumulate retirement benefits through a civilian employer. They are frequently relocated, uprooting them from their friends and community, and the moves diminish the employment and earning opportunities of the military spouse. In addition, the calculation of basic pay in determining retirement benefits favors reservists. Basic pay for a retired reservist is the value of basic pay in effect when the reservist turns 60, not the value of basic pay in effect when the reservist separated plus the cost of living adjustment to age 60.

Reserve deployments increased during the 1990s and rose further during the operations in Iraq and Afghanistan. However, increasing the generosity of reserve retirement benefits appears to be an inefficient, poorly targeted, and unfair way of compensating for the higher burden of deployment. It is inefficient because it is more costly in the current period for the government to provide a benefit in a future period than that benefit is worth to the reservist in the current period. Instead, it would be more efficient (and less costly) to offer cash pay today. Using retirement benefits to address the stress from greater deployment is not well targeted, because more-generous retirement benefits potentially reward all reservists, including those with little or no deployment.

Flexibility in managing personnel allows the services to tailor military careers to obtain the best retention profile in each occupational area. But the reserve retirement benefit system, like the active system, has operated to create uniformity in careers, retention, and incentives for performance. In contrast, increasing current pay through special and incentive payments and revising the personnel management system can increase flexibility in managing personnel. A combination of special and incentive payments and retirement-system reform can be used to allow greater overall flexibility in force management.

Toward Broader Reform of Military Retirement

Our analysis of reserve retirement reform also considered the broader policy context—and specifically, reform proposals to change the active retirement system. Because both the active and reserve retirement systems have remained virtually unchanged since the 1940s despite frequent calls for change, we also sought to identify the obstacles to reform and the factors that might improve the prospects for successful changes to the active and reserve retirement systems.

Past commissions and study groups have been appointed to examine military compensation and AC retirement benefits. The issues they addressed are the cost of AC benefits, the inequity of paying benefits only to those who serve 20 years, the lack of management flexibility, and the similarity in career length produced by the 20-year cliff-vesting rule. These issues are increasingly relevant to the reserves. Achieving a compensation system that supports the seamless integration of reservists called to active duty will require that reserve retirement reform be coordinated with active reform, although the resulting systems will not necessarily be identical.

In addition to cost, another key obstacle to reform is a lack of consensus for change among the services, the U.S. Department of Defense (DoD), service members, and retirees. The AC and RC retirement systems have delivered a steady supply of experienced manpower. The services seem to have adapted their manning requirements to this retention. Perhaps tasks and jobs have been designed to accommodate the retention profile produced by the compensation system, promotion system, and cliff-vesting at 20 YOS. Past study groups have not demonstrated that there is an excess supply of senior personnel, nor provided evidence of the potential gains in defense capability and readiness from greater flexibility (which is not to deny the existence of such gains). Lacking compelling evidence of its benefits, the services have not been strong advocates of retirement reform. Veterans and retiree groups have also not advocated reform, possibly fearing that it would open the door to benefit cuts and broken trust with service members. A lack of consensus for reform might also come from differing views on the objectives of the retirement systems. Advocates of change may emphasize the role of the retirement system as a policy instrument for force management, while advocates of the current system may emphasize its role as a reward for a long career of service, a benefit to help transition to the civilian economy, and a means of securing the retention of career personnel.

To garner insight into the factors affecting whether reform will be sought and into strategies to increase the chance of successful reform, we adapted an economic theory of compensation reform under political constraints to the issue of retirement reform. Based on the model, here are some key elements for developing a strategy to address the obstacles to reform:

- Identify the chief constituents involved in the retirement reform process, e.g., service members and DoD leadership. Define their different objectives of reform.
- Assess the gains and losses of reform to the constituents. For example, provide the services with information on how retirement reform would improve defense capability.
- Recognize that buy-offs and other factors that make reform more palatable are necessary to make the reform politically feasible. Buy-offs may be paid up to the point where they do not outweigh the improvement in defense capability or cost-effectiveness associated with the reform.
- A menu of reform choices can improve the feasibility of reform and reduce the cost of buy-offs.

Regarding the menu of choices, we discuss a case study of a government retirement system that was highly effective; namely, the transition of federal employees in the 1980s from the civil service retirement system (CSRS) to the federal employees retirement system (FERS). Based on the CSRS-FERS case study, we add the following lessons about effective reform:

- Provide a menu of choices that includes the existing retirement system as well as the new retirement system.
- Provide multiple enrollment opportunities to switch to the new retirement system, so the menu is an ongoing choice.
- Design the new retirement system to be more generous on average for new and junior employees but less generous for existing senior employees, thereby minimizing the amount of switching on average. Alternatively, "grandfather in" incumbent employees.
- Design the new system so that it is portable to other jobs.

Acknowledgments

We thank Richard Krimmer, former Director of Military Personnel Programs, Office of the Assistant Secretary of Defense for Reserve Affairs, and Captain Michael Price, former Assistant Director of Military Personnel (Compensation) and Coast Guard Liaison to the Office of the Assistant Secretary of Defense for Reserve Affairs. We are grateful to Harvey Johnston, Defense Manpower Data Center (DMDC), who arranged for the administrative data files used in the analysis; Dr. Tim Elig, formerly of the DMDC, who arranged for the Status of Forces of Reserve Personnel survey data; and Scott Seggerman, DMDC, and Virginia Hyland, Office of the Assistant Secretary of Defense for Reserve Affairs, who provided information on reserve component deployments.

At RAND, Daniel Clendenning and Tina Panis programmed early versions of the model; John Christian, Phoenix Do, Serhii Ilchuk, and Neda Farzinia provided research assistance; Beth Roth processed the active/reserve database and developed analysis files; and Mark Totten handled programming and analysis tasks for the reserve surveys. The initial version of this research benefited from reviews by Jacob Klerman and Michael Mattock, and Michael participated in the substantial reworking of the model and analysis and has become a co-author. We also benefited from the comments of Eric Peltz at RAND.

We appreciate support received from officials in the Office of the Undersecretary for Personnel and Readiness. These individuals and their affiliation at the time of the initial analysis include Saul Pleeter, Office of Compensation; Chuck Witschonke, Associate Director of the Office of Compensation; John Winkler, Principal Deputy Assistant Secretary of Defense for Reserve Affairs; Wayne Sproul, Principal Director, Manpower and Personnel, Office of Secretary of Defense for Reserve Affairs; and Tom Bush, who succeeded Mr. Sproul as Principal Director, Manpower and Personnel and became Director, Program Integration Office, also in Reserve Affairs (Manpower and Personnel).

Portions of the report draw on a 2006 RAND technical report, *Reserve Retirement Reform: A Viewpoint on Recent Congressional Proposals*, by Beth Asch, James Hosek, and David Loughran, as well as a 2012 report, *Reserve Participation and Cost Under a New Approach to Reserve Compensation,* by Michael Mattock, James Hosek, and Beth Asch.

Abbreviations

AC	active component
BAH	basic allowance for housing
BAS	basic allowance for subsistence
BHHH	Berndt, Hall, Hall, and Hausman
CPI	Consumer Price Index
CSRS	Civil Service Retirement System
DB	defined benefit
DBB	Defense Business Board
DC	defined contribution
DMDC	Defense Manpower Data Center
DoD	U.S. Department of Defense
DRM	dynamic retention model
DSB	Defense Science Board
FERS	Federal Employees Retirement System
FY	fiscal year
IDT	inactive-duty training
PCS	permanent change of station
QRMC	Quadrennial Review of Military Compensation
RC	reserve component
RC/T	Reserve Component/transit
REDUX	1986 revision to military retirement system
RMC	regular military compensation
SSN	Social Security Number
TSP	thrift savings plan
WEX	Work Experience File
YCS	years of creditable service
YOS	years of service

CHAPTER ONE

Introduction

Selected reserve forces have been deployed extensively throughout the military operations in Iraq and Afghanistan and in military operations in the 1990s. The nation's growing reliance on the reserves has led to heightened interest in reserve manpower policy, touching on reserve training, reserve unit cost, personnel turnover in reserve units scheduled for deployment, reserve family support, health benefits for activated reservists, the well-being of children and caregivers of deployed reservists, and others. The adequacy of reserve compensation and reserve retirement benefits has also come into question.

The reserve component (RC) and active component (AC) compensation systems differ in significant ways, and considerable attention has been paid to those differences. Like the active system, the reserve system rewards members who complete 20 years of service (YOS) with a retirement annuity based on years of service and basic pay. But some argue that the current system is inequitable because a reservist who completes 20 creditable YOS may not draw retirement benefits immediately after retiring from the military, while an AC retiree may do so. Reservists are typically on duty part of a year and receive credit for that part. The retirement benefit computation adjusts for the fact that reservists serve part-time unless they are activated or have full-time assignments with AC units, in which case they accumulate credit at a full-time rate like AC personnel. Recognizing that this adjustment is already built into the system, critics contend that reservists should be treated like actives and receive an immediate annuity. In 2008, Congress passed legislation that permits reservists to start retirement benefits three months sooner for every 90 days of deployment occurring after January 28, 2008. But differences in the compensation systems remain, and they continue to be the focus of policy debate and rancor among military members and veterans.

Evaluation of the desirability of proposals for reform of the military compensation system, including changes to the RC retirement system, requires an empirically based modeling capability to quantitatively assess the cost and force-management implications of the proposals in terms of their effects on active and reserve retention. It also requires an understanding of the broader policy context, including the proposals' effects on military members and the political feasibility of reform of the compensation system.

Over the past several decades, RAND has developed an analytic capability called the dynamic retention model, or DRM, to make quantitative assessments of the force-management and cost effects of military compensation reform proposals. The capability was most recently used to assess RC compensation reform proposals for the 11th Quadrennial Review of Military Compensation.

This document considers reserve retirement reform. We use the DRM to assess three past congressional proposals from the 108th and 109th Congresses that aimed to increase reserve retirement benefit generosity by decreasing the age at which a reservist could draw retirement benefits. The bills differed in their approach:

- H.R. 331 (108th Congress) provided benefits immediately on retiring from the military with 20 or more YOS, regardless of one's age (we refer to this as the "immediate annuity" proposal).
- S. 32, S. 38/H.R. 1169, H.R. 783 (109th Congress) decreased the age of first receipt from age 60 to age 55 ("age 55").
- S. 337/H.R. 558 (109th Congress) introduced a sliding scale that would have decreased age of first receipt by one year for every two years served beyond 20 years: A reservist could retire at age 60 with 20 YOS; at age 59 with 22 YOS; and so forth to age 53 with 34 YOS ("sliding-scale").

None of the proposals were actually legislated, but they represent the types of changes that continue to be advocated by those who desire a closer alignment of the RC and AC retirement systems. Furthermore, the debate surrounding them led to the aforementioned 2008 alteration of the RC retirement age (three months early for every 90 days of deployment).

From a broader perspective, the bills reflect considerations of competitiveness, fairness, and equity that underlie policy change. Higher retirement benefits are a way of improving the competitiveness of reserve compensation in view of plans for continued use of reserve forces along with active forces in future military operations. Also, higher benefits are fair if the reserves bear more of the defense burden than before, and higher benefits would bring them more in line with AC retirement benefits.

Efficiency and equity are good reasons to propose changes to the reserve retirement benefit system, but analysis is required to determine by how much a proposed policy would increase reserve manpower supply and at what cost. Our research provides such analysis.

We analyze the proposals above to determine their likely effect on active retention, reserve affiliation after leaving an active component, reserve retention, and reserve retirement. We present a unified model of these behaviors, fit the model to actual data, and use the estimated model for policy simulations of the proposed reserve retirement reforms. We presented early results of the analysis to the Office of the Secretary of Defense for Personnel and Readiness and at a conference on the changing roles for the reserve components and implications for compensation and personnel policy (Asch and Hosek, 2008). But we chose not to publish our complete analysis at that time because we were still refining our estimation code and policy simulation modules in the midst of policy analysis for the 10th Quadrennial Review of Military Compensation. Shortly after the 10th QRMC we began analysis for the 11th QRMC on possible changes to current RC compensation, and this allowed further development and refinement of our modeling. This document, then, takes advantage of these advances and revises the policy simulations of the congressional proposals. For this, we use our newer parameter estimates, which are precise and stable. We find that the updated results are consistent with the earlier results and the implications are the same.

We go beyond policy simulations to discuss issues related to reserve retirement reform. These include a description of the context of active and reserve service, which is useful for thinking about the equitability of RC and AC retirement benefits and about reserve force-management flexibility. Because of the relationship between the AC and RC and their respective compensation systems, we also relate reserve retirement reform to active retirement reform and review past proposals for active reform. Finally, because no major changes to the AC or RC retirement systems have occurred since the 1940s, despite decades of debate, we also consider the obstacles to such changes and discuss factors that might increase the chance of successful reform.

Chapter Two describes our model, data, estimates, and goodness of fit, and Chapter Three presents policy simulation results. Chapter Four discusses additional issues related to RC retirement reform, while Chapter Five extends the discussion to AC retirement reform. Chapter Six considers obstacles to change, and Chapter Seven has conclusions and policy implications. We include the technical details of our model and analysis in the appendixes.

CHAPTER TWO

Model, Data, and Estimates

We have found that many audiences are unfamiliar with the dynamic programming approach we use, and in this chapter we describe our model, data, estimation technique, and fit. Some readers, however, may want to skip to the next chapter, which presents the policy analysis of the congressional proposals.

Model

Our model is a stochastic dynamic programming model of AC retention and possible subsequent RC participation. The model focuses on individuals who begin their military service in an active component.[1] It treats the individual as the decisionmaker, assumes he or she faces future opportunities that are partly known and partly uncertain, and allows the individual to reoptimize in each period depending on his or her state in that period and its realizations of previously uncertain factors. An individual's status may be active, reserve, or civilian, and in any of these statuses the individual's state is defined by years of active service, years of reserve participation, and total years. The difference between total years and years of active plus reserve service equals years of civilian work experience. The model is both forward-looking and accounts for past events and decisions that have brought the individual to the current state. Current and future opportunities, including retirement benefits, affect the current decision given the current state.

The model builds on seminal dynamic programming research at RAND from the 1950s by Richard Bellman, who formulated stochastic dynamic programming, and from the 1970s and 1980s by Glenn Gotz and John McCall (1984), who applied it to military retention and were the first to estimate a dynamic retention model. Our

[1] Appendix B has a technical presentation of the model. The model specification can also be found in Asch and Hosek (2008). As mentioned in Chapter One, we updated our analysis by drawing from the estimates we produced for the 11th QRMC. Therefore, our description of the estimation approach, simulation approach, and model fit draw from Mattock et al. (2012).

model extends the Gotz-McCall model to include the reserves and to allow the reserve and civilian alternatives to be nested.

Dynamic programming models of retention differ from traditional approaches that focus on retention at an isolated decision point, such as the end of the first or second term. The dynamic programming approach describes an individual's retention decisions over the lifecycle. It assumes individuals differ in their "tastes" for active and reserve service, and individuals with different tastes may respond differently to the same policy. Also, each period may have random outcomes such as a good or bad assignment, location, or job opportunity. When applying the model to data, the parameters to be estimated include those of the taste and shock distributions. The model also embeds details of the compensation and retirement systems. When the model parameters have been estimated, the model can be used to simulate alternative military compensation and retirement policies.

The model assumes that individuals behave in a rational, time-consistent manner. The estimated model fits data on career retention well, as shown later in this chapter. This may result in part from the general stability of military compensation and promotion that enables service members to form accurate expectations of their military pay. The structure of basic pay and retirement benefits has been quite stable for nearly 60 years, and occasional changes—such as targeting higher basic pay increases on more-experienced and faster-promoted members or increasing pay in the early years of service, which was done to launch the all-volunteer force—have been infrequent and widely publicized. The promotion system also has been generally stable.

The assumption of time-consistent behavior is reasonable. Behavior is time-inconsistent when, for example, it is in the individual's longer-term interest to take an action but the individual chooses not to act in the short term (Laibson, 1997). In civilian retirement plans, a concern is that individuals delay enrollment even though it is in their interest to enroll as early as possible. But in the military retirement system and the proposals we consider, enrollment is automatic and retirement contributions are made by DoD.

Another model of military retention, the annualized cost of leaving model, is known not to be time-consistent. For instance, the model might predict in a member's eighth year of service that he or she will leave in their 14th year, but when that time arrives the member might stay. The dynamic programming approach avoids this type of inconsistency by permitting reoptimization in each period and through a structure that builds in forward-looking behavior.

The model focuses on individuals without prior military service who join an active component. An AC member has three alternatives in each period: stay in the AC, leave and join the RC, or leave and become a civilian. Reservists and civilians are assumed to hold civilian jobs, and reservists belong to a component of the Selected Reserve, i.e., Army National Guard, Army Reserve, Navy Reserve, Marine Corps Reserve, Air National Guard, and Air Force Reserve. Although a person leaving the AC might join any of the Selected Reserve components, we assume he or she would join a unit in the same branch of service. Also, we pool the Army National Guard and the Army Reserve into one group, likewise the Air National Guard and the Air Force Reserve. For example, someone leaving the AC Army might join the "Army reserve," meaning either the Army National Guard or the Army Reserve. We assume that after separating from the AC, a member may not reenter the AC but can move back and forth between reservist and civilian statuses. The model does not include the active accession decision, although this could be added in future work.[2] Also, the modeling approach can be extended to individuals who begin their career in the RC.

We discuss the reserve retention decision first and the active retention decision second, then conclude this section with a summary of the model.

Reserve Decisionmaking

At the end of each period, the reservist compares the value of staying in the reserves with the value of leaving to be a civilian and chooses the alternative with the higher expected value. Today's decision can affect tomorrow's opportunities, so the expected value computation is more complex than simply multiplying together the probability of an event and the outcome associated with that event and summing. The reservist considers all future paths attainable from his current status and state and, in making the optimal decision today, is assumed to make the optimal decision in each future period. Status is defined by whether the member is active, reserve, or civilian. State is defined by the member's career history and specifically by the number of active years, reserve years, and total years accumulated up to a given point in time.

Staying in the reserves gives the individual the option in the next period of choosing to stay in the reserves or leave, and the value of the option takes into account the added reserve experience from the current period. Another year of reserve experience leads to higher military pay, as well as more points and creditable years toward retirement. Because the model assumes that the reservist is employed in a civilian job, another year in the reserves implies another year of civilian experience, so the civilian wage increases too.[3] If the reservist leaves the reserves, military experience does not increase but civilian experience does and so does the civilian wage. A civilian can reenter the reserves. A positive reserve shock or negative civilian shock makes entering the reserves more attractive.[4]

---

2 For instance, Hosek, Mattock, et al. (2004) include active component enlistment in their dynamic retention model.

3 Staying in the reserves might alter civilian options upon leaving the reserves; e.g., the reservist might acquire skills that are valuable on civilian jobs. We do not model this possibility.

4 "Shock" refers to a random draw from a distribution representing events that can affect the individual's perceived value of a choice, such as the choice to continue in the active component, or leave to become a civilian or reservist. This approach is used because such events should be recognized in the decision, yet most data sets, including ours, do not contain variables for the various types of events that could affect these values. Such events

To compute the expected career path from a given current state, the individual evaluates the payoff to all possible career paths and then weights each path by the probability it will be chosen, where the probability depends on the expected payoffs (i.e., the values of the value functions). To obtain the expected payoffs, the dynamic programming model is solved by backward recursion from the last decision period; the model is a finite-period dynamic program.

In the last period in which the individual may serve in the reserves, there is a terminal condition; in our model, when the individual has a total of 30 years of active plus reserve service, no further military service is allowed and the individual must become a civilian. The model assumes a 40-year work life from ages 20 to 60, and there are no decisions after age 60.

In the second-to-last period, the individual can opt to be a reservist or a civilian and will be in the chosen state in the final period, choosing "reservist" or "civilian" as the best option based on the information available at the time. However, when the individual is in an *earlier* period looking ahead to the second-to-last period and does not know what the shocks will be in that period, both "reservist" and "civilian" have a chance of being the best choice.[5] This approach captures the value of being allowed to choose between the two options when the time comes. Given this assessment, the individual does not commit today to stay or leave in the second-to-last period but instead constructs rules for making the optimal choice conditional on his state at that time and the range of shocks that support "civilian" or "reservist."

Thus, in each period, the individual reasons both backward and forward. Given his state at the time, he reasons forward about the states he might transition to, then works backward for each possible path from the terminal period to construct rules for making the optimal choice in each upcoming period. The rules allow the individual to determine the probability of being in each possible state in each future period, and the solution continues backward to the current period and state. In making his decision, the individual also considers discounted values of upcoming periods, current pay, and shocks in the "reservist" and "civilian" alternatives.

The probability of a particular choice in a given period can be multiplied by choice probabilities in other periods to give the probability of a career path, i.e., a particular sequence of "reservist" and "civilian" statuses and states.

---

might include assignments, location, missions, leadership, individual and unit preparedness, family exigencies, job opportunities in the civilian world, illness, and so forth. Although in reality there may be many such events in any period, we use a single shock (random draw) to represent their net effect. Thus, the random draw can be positive or negative, big or small.

[5] Since either could be best, the individual allows for this by taking the expected value of the maximum, which is the sum of the expected value of "reservist" (given that it is greater than "civilian") multiplied by the probability that "reservist" is greater than "civilian," plus the expected value of "civilian" (given that it is greater than "reservist") multiplied by the probability that "civilian" is greater than "reservist."

The model assumes the individual's taste for reserve service is constant. Those with a higher taste for reserve service will have a higher probability of a career with many years of reserve service.

Another factor affecting the reserve retention decision is the personal discount rate. A high personal discount rate implies that future compensation has a much lower value to the individual than current compensation. At a 10 percent personal discount rate, the current value of $1 is $.625 if paid years 5 from now, $.39 if paid 10 years from now, and $.15 if paid 20 years from now. We estimate the personal discount factor (1/(1 + personal discount rate)) for enlisted personnel and officers, by service.

Active Decisionmaking

An AC member chooses between active, reserve, and civilian statuses. Those who leave active duty may not return to it, while those who choose to stay can serve up to a 40-year career.

Each individual has some taste for active service, and the active and reserve tastes are assumed to have a joint distribution. The joint distribution has five parameters of interest: the means and variances of active and reserve tastes and the correlation between them.

Those with a high taste for active service place a high value on continued service in the active component, those with a high taste for reserve service place a high value on affiliating with the reserves, and those with low active and reserve tastes are likely to become civilians. For those opting to be a reservist or a civilian, the preceding discussion describes their decisionmaking once they leave the active component. Similar to the discussion above, the value of staying in the active component is computed by evaluating each future career path starting from the current active state and weighting by the probability the path is chosen. Again, the expectations recognize that the future is uncertain, and the individual takes the expected value of the maximum of the three choices (active, reservist, and civilian).

However, from the perspective of an individual in the AC, the model nests the "reserve" and "civilian" alternatives. (The nest structure is explained in Appendix B.) Most reservists also hold a civilian job, and nesting allows for a shock to both "civilian" and "reservist." In addition to a common, nest-level shock, "civilian" and "reservist" are each subject to their own shock (as discussed above). In the nesting approach, the individual's thought experiment is to choose the maximum of reserve versus civilian, and then to choose between active and this maximum. While the individual observes all the shocks for the current period, he or she does not know the shocks in future years. However, the individual is assumed to know the distributions from which shocks are drawn and use this information to assess the expected maximum value of the choices in each future period.

Model to Estimation

The model portrays AC and RC retention choices in a stochastic dynamic program over a finite horizon, embeds uncertainty about future conditions (shocks), and permits tastes for active and reserve service to differ across individuals. We as researchers do not observe these tastes, but we assume they follow a certain joint probability distribution.[6] The model builds in uncertainty in the form of random shocks, and we assume the shocks follow certain probability distributions. We use the model's structure, the distributional assumptions about tastes and shocks, and information about military pay, military retirement benefits, and civilian pay to derive expressions for the probability of a military career represented by AC retention and possible RC participation by period. For instance, a person might serve five years in the AC, then three years in the selected reserve, then work as a civilian for two years followed by a year in the RC, finally settling as a civilian in all remaining periods. Such a career would look like this: {A,A,A,A,A,R,R,R,C,C,R,C,C,...,C}. We have longitudinal data and create probability expressions for each person's career. These career probabilities are multiplied together to obtain a maximum likelihood expression for the entire sample, and this expression is maximized with respect to the model parameters to obtain parameter estimates.

Data

Work Experience File Data

Our main data file is the Work Experience File (WEX). The WEX contains person-specific longitudinal records of active and reserve service.[7] The Defense Manpower Data Center (DMDC) creates WEX data from the active-duty master file and the RC common personnel data system file. DMDC uses these files to build a snapshot of all personnel for each reporting period that includes demographic and work experience information. Demographic information includes scrambled Social Security number (SSN), name, service/component, reserve category code, pay grade, date of birth, years of service computed from pay entry base date, sex, and language skills. To maintain the file, DMDC compares data for the current and previous periods and creates three types of records: a gain record, a loss record, and a change record. A gain record is created when a service member's SSN is not in the previous period but is in the current one. A loss record is created when an SSN appears in the previous period but not in the current one. When a loss occurs, all related work experience records are moved to a loss

---

6 The AC taste distribution evolves as years in the AC increase, as those staying in an active component are likely to have higher tastes than those leaving. Similarly, the RC taste distribution evolves as years of RC participation increase.

7 WEX is used primarily for production of Verification of Military Training and Experience DD Form 2586 documents.

file; these are retrieved when an individual reenters service; i.e., when a gain occurs. A change record is created when there is a change in any of seven variables: service/component, pay grade, reserve category code, primary service occupation code, secondary service occupation code, duty service occupation code, and unit identification code. The WEX record also includes a member's age and gender.

WEX data begin with service members in the AC or RC on or after September 30, 1990. Our analysis file includes AC non-prior-service entrants in 1990 and 1991 followed through 2010, providing 21 years of data on 1990 entrants and 20 years on 1991 entrants. For each AC component, we drew samples of 25,000 individuals who entered the component in fiscal year (FY) 1990–1991, constructed each service member's history of AC and RC participation, and used these records in estimating the model. We use WEX variables to identify an individual's component and branch of service (e.g., AC Army, RC Army Reserve) by year from the date of entry onward. We use pay entry base date and component/branch in counting years of AC service and years of RC participation following AC service.

We construct samples for each service and for enlisted personnel and officers. In constructing the officer samples, we exclude medical personnel and members of the legal and chaplain corps because their career patterns differ markedly from those of the rest of the officer corps. Analysis of retention for these personnel needs to be conducted separately. For a similar reason, for Air Force officers, we exclude rated pilots. A dynamic programming model of Air Force officer pilot retention can be found in Mattock and Arkes (2007).

Basic Pay, Regular Military Compensation, and Retirement Benefits

AC pay, RC pay, and civilian pay are averages based on the individual's years of AC, RC, and total experience, respectively. AC and RC pay are also related to military retirement benefits, as discussed below. We use 2007 military pay tables, but because military pay tables have been fairly stable over time, with few changes to their structure,[8] we do not expect our results to be sensitive to the choice of year.

Annual military pay for AC members is represented by regular military compensation (RMC) for FY 2007 given years of active service. RMC accounts on average for over 90 percent of the cash pay received by active duty personnel. It is the sum of basic pay, basic allowance for subsistence (BAS), basic allowance for housing (BAH), and the federal tax saved because the allowances are not taxed. We compute AC pay by year of service for enlisted and officers by branch of service. RMC in general depends on AC years of service, pay grade, and dependents status, but pay grade and dependents status are omitted from our model. This means that we do not explicitly include prob-

---

[8] An exception was the structural adjustment to the basic pay table in FY 2000 that gave larger increases to mid-career personnel who had reached their pay grade relatively quickly (after fewer years of service). A second exception was the expansion of the basic allowance for housing (BAH), which increased in real value between FY 2000 and FY 2005.

abilities of promotion, up-or-out rules, marriage, or divorce/separation.[9] The AC pay variable at a given year of service is the average RMC at that year, where the average is taken over the number of service members in each pay grade at that year and whether they have dependents. Information on grade distribution and dependents comes from the defense budget estimates for FY 2007 (Office of the Under Secretary of Defense, 2006). We obtain a rough estimate of the tax advantage by computing the percentage of AC RMC that is attributable to it and applying that percentage (roughly 6 percent) to the sum of basic pay, BAS, and BAH for AC members. While greater precision in estimating the tax advantage would improve our estimates of AC RMC, our purpose is not to provide such an estimate per se, but to provide an input to our model. We believe that our parameter estimates are not sensitive to our approach to computing the tax advantage.

RC members are paid differently than AC members although the same pay tables are used. Reservists who are drilling but not on active duty receive subsistence allowance for their two drilling days per month and do not receive a housing allowance. Reservists on active-duty training receive rations and housing in kind during the two weeks of training and receive either a partial housing allowance or a rate applied for married members, unless they are housed in contract housing off-base.

RC pay is based on years of AC service and years of RC participation. We average it over pay grade and dependents status using RC strength information from the 2007 *Official Guard and Reserve Manpower Strengths and Statistics Report* (Office of the Assistant Secretary of Defense, Reserve Affairs, undated). Reserve pay in a year is calculated as the sum of drill pay for four drills per month, 12 times a year, plus pay for 14 days of active-duty training, typically done in the summer. Drill pay is 1/30 of monthly basic pay for each drill period, or 4/30 per weekend. During each day of active-duty training, the reservist receives basic pay plus BAS. Single members receive BAH for a service member without dependents, while married members receive BAH for a service member with dependents. In our calculation, RC members receive BAH RC/T (Reserve Component/transit), a housing allowance for certain circumstances, including being on active duty less than 30 days. Given years of service and grade, we compute a reservist's annual pay as:

12 x weekend drill pay + 14 x
((BAS + daily basic pay) + % married x BAH RC/T for those with dependents +
% single x % on base x BAH RC/T for those without dependents) + tax advantage.

[9] Pay grades, promotion probabilities, and up-or-out rules were included in our model for the 10th QRMC, but they have been omitted here because the RC compensation changes under consideration are not aimed at changing promotion speed or up-or-out rules, and the model runs faster without these features. That said, insofar as personnel policies such as these influence the retention patterns we observe in the data, the parameter estimates we get using these data will indirectly incorporate the effects of these policies.

To incorporate the tax advantage, we use the same adjustment as for AC annual pay, 6 percent. Some reservists receive special and incentive pay such as bonuses, but these are not included. Also, the model does not address the activation and deployment of reservists.[10]

The reserve retirement benefit formula and the High-3 active duty retirement formula are programmed into our model. These formulas are described in Appendix A. In short, the AC retirement system vests at 20 YOS in an immediate annuity that is defined by a formula. The formula is 2.5 percent times the average of the member's highest 36 months of basic pay prior to retirement, or "High-3" years, multiplied by years of service. Eligibility for RC retirement benefits requires 20 years of creditable service. Years of creditable service include AC years plus years of RC participation where the reservist earns at least 50 points. A reservist receives 15 points for affiliating with a selected reserve unit, plus one point per drill and one point per day of active-duty training. For example, a reservist who attends all drills and active-duty training might accumulate 77 points (15 + 12 x 4 + 14) and therefore will have a creditable year. We assume an RC participant accumulates 75 points per year. Unlike AC retirement benefits, which start as soon as the AC member retires from service, RC retirement benefits begin at age 60.[11] The formula for RC retirement benefits is the same as that for AC retirement benefits, with the proviso that RC retirement points are converted into years of service (for the purpose of retirement) by dividing total points by 360. A year of AC service counts as a full year. Reservists who qualify for reserve retirement benefits can transfer to the "retired reserve," which means that their High-3 pay is based on the basic pay table in place on their 60th birthday, and their basic pay is based on their pay grade and years in grade, where the latter include years in the retired reserve.[12]

Civilian Earnings

For enlisted personnel, we use the 2007 median wage for full-time male workers with an associate's degree. For officers, we use the 2007 median wage for full-time male workers with a bachelor's degree or more. The data are from the U.S. Census Bureau. Civilian pay varies by total work experience, defined as the sum of active years, reserve

---

[10] The pay of approximately 85 percent of activated reservists is higher than the sum of their reserve pay and civilian earnings when not activated (Loughran et al., 2006).

[11] If the RC member has been deployed in the period beginning on January 28, 2008, retirement age is decreased by three months for every 90 consecutive days of deployment. This change is not included in our model because the model does not include deployment.

[12] In addition, military retirees (including reserve retirees receiving retired pay) are eligible to receive health care through TRICARE for the remainder of their lives, as can their spouses, and coverage continues for the spouse if the retiree dies and she or he does not remarry. "Gray area" retirees (i.e., members of the retired reserve who are not drawing retired pay) may purchase TRICARE coverage under the TRICARE Retired Reserve program until they become eligible for TRICARE. We do not model the health benefit, however.

years, and civilian years since age 20, but here does not vary by other factors such as years since leaving active duty.

Estimation

Thus the basic model has eight parameters: AC mean taste, RC mean taste, AC taste variance, RC taste variance, AC/RC taste covariance (or correlation), shape parameter for the RC/C nest shock, shape parameter for the RC and civilian shocks, and personal discount factor. The density for the taste distribution is bivariate normal. The shock densities are extreme-value with mode zero, thus two shape parameters need to be estimated: one for the AC/C nest and one specific to the civilian/reserve alternatives in the nest. In addition, we estimate parameters for switching costs. These reflect the cost associated with switching from one state to another.

The model is estimated using maximum likelihood methods. Writing down the likelihood function requires us to compute the probability of a given career, consisting of a sequence of active, civilian, and reserve states. This computation, in turn, requires us to compute the probability of choosing each alternative in each time period. Given our assumption of an extreme-value distribution for the shock terms, we can solve the dynamic program given values for active and reserve taste. The solution gives closed-form solutions for the probability of choosing each of the two or three alternatives available at any given time, and as mentioned we use these to construct a career probability for each individual. The expression for the career probability implicitly depends on the parameters to be estimated, e.g., mean active taste, mean reserve taste, discount rate, and so forth. Because tastes are not known at the individual level, we numerically integrate out heterogeneity in taste.[13] Numerical optimization is done using a standard Berndt, Hall, Hall, and Hausman (BHHH) hill-climbing algorithm (Berndt et al., 1974).

Standard errors are computed using numerical differentiation of the likelihood function at the parameter estimates to produce the matrix of second derivatives, the Hessian matrix. The standard errors are computed using the square root of the absolute values of the diagonal of the inverse of the Hessian. We report the parameter estimates and standard errors by service for officers and enlisted in Appendix C.

To judge goodness of fit, we use the parameter estimates in conducting a simulation of AC and RC retention patterns by year of service. These simulations are compared to the actual data to assess the extent to which the model predicts actual behavior, i.e., to assess model fit. We also simulate AC and RC retention under the alternative

---

[13] For trial values of the taste distribution parameters, possible tastes for the individual are drawn from the distribution. For each taste draw, the career probability expression is evaluated and an average of those probabilities is taken, where the weight on a probability depends on the probability of the drawn taste.

policies under consideration. The next section describes our simulation approach and model fit. Results of simulating the policy proposals are presented in the next chapter.

Approach to Simulation

To simulate retention behavior, we first create a synthetic population of 10,000 individuals entering active duty by randomly drawing tastes from the estimated AC/RC taste distribution. Each pair of AC and RC taste draws represents an individual entering active duty. We also draw shocks for each year for each synthetic individual from the shock distributions. We assume that the synthetic individuals follow the logic of the model, and we specify the compensation policy for the simulation. We simulate behavior under the current compensation policy, the baseline, and then simulate it under the policy alternatives. The simulations produce a 40-year record of AC retention and RC participation for each member of the synthetic population under each compensation policy.

We use the data sets of simulated behavior to tabulate AC and RC retention and, along with information on compensation, to compute policy cost. The simulation outputs include graphs of AC retention by year of service and RC participation by year of active-plus-reserve service, as well as tabulations for AC and RC of force size, current pay cost, retirement cost, and total cost (the sum of current and retirement cost).

Under the assumption of a steady state, the AC force size of the simulated population is the count of individuals present in each year up to year 40. This count is scaled up to AC force size for each service. For example, the simulation results for the Army enlisted force are scaled up to 458,220, a number equal to the 2009 size of the active-duty enlisted Army. The corresponding figures are 272,208 for the Navy, 263,351 for the Air Force, and 182,366 for the Marine Corps. Similarly, we scale up the simulation results for officers based on the size of these forces in 2009: 90,795 for the Army, 52,031 for the Navy, 65,496 for the Air Force, and 20,709 for the Marine Corps.[14]

RC force size is based on the count of simulated individuals participating in the RC at each year of service, given scaling up the AC force to its force size. As mentioned, RC years of service are based on the number of active years plus reserve years.[15]

---

[14] Our analysis of Air Force officer retention is for nonrated officers. We exclude pilots because their career pattern differs substantially from that of nonpilots. Because nonrated officers are about half the force, we scale the officer results to 0.5 x 65,496 or 32,748.

[15] As an example of this count, consider someone who over the course of 40 years (ages 20 to 60) had 5 years of AC and 5 years of RC service. This individual would be present in the RC at 6 YOS (5 + 1), 7 (5 + 2), 8, 9, and 10. (Participation in the RC could have occurred in nonconsecutive calendar years.) In each of these years, the individual would be counted in the steady-state RC force. Because everyone begins in the AC, the smallest RC YOS entry is 2 (1 + 1).

Simulation of Cost

AC current cost equals RMC at each year of service multiplied by the number in AC in that year, summed over all years. AC retirement cost is computed as a normal cost percentage of the basic pay bill for the AC force, consistent with the practice of the DoD Actuary. This gives an amount, an accrual charge, sufficient to cover the retirement liability of AC service members who retire from the AC plus a portion of the retirement liability of AC members who retire from the RC. AC current and retirement costs are also scaled up.

RC current cost is the product of RC pay by year of service, summed across years. RC retirement cost is based on the reserve retirement liability for the reserve force less the funding credited to the reserve retirement account from the accrual charges made during its AC service. This also follows the practice of the DoD Actuary. Specifically, the amount transferred from the AC retirement fund to the RC retirement fund is based on calculations involving the number of AC members who leave at each year of AC service and subsequently qualify for RC retirement.[16]

Model Fit

Figures 2.1 and 2.2 show the model fit for enlisted personnel and officers by branch of service. Small circles show actual retention and a line shows simulated retention. The simulations, which are based on the current compensation system,[17] are quite close to the actual data, providing evidence that the model fits the data well.

We next turn to our analysis of the congressional proposals for reserve retirement reform.

---

16 The actuarial calculation is made for those leaving AC by AC year of service. For example, consider 100 AC service members in the 10th year of service: Suppose that 80 later qualify for AC retirement, while six leave the AC at the end of 10 YOS and later qualify for RC retirement. With our simulated population, we can determine the years of service and pay at which they retire, and from survival tables we know how long they are likely to live. This allows us to compute the total retirement liability of RC retirees. Our understanding is that 6 percent of the AC accrual charges during AC years 1 through 10 are transferred to the RC retirement fund on behalf of the six individuals who will retire from the RC.

17 This system has remained in place, although with some changes, over the 20-year period represented in our data, including a change in FY 2000 to allow members who entered after August 1986 to choose at 15 YOS between the high-three retirement system and the REDUX retirement system with a bonus. In the late 1990s, military pay lagged civilian pay and Congress mandated a catch-up basic pay increase for FY 2000 and higher-than-usual basic pay increases over the next six years. Higher-than-usual increases in fact continued through FY 2010. BAH was increased during FY 2000–2005, and bonuses were heavily used in 2005–2008 to sustain recruiting and retention. Military retirement benefits and eligibility rules did not change but TRICARE for Life was implemented, giving military retirees continued eligibility for TRICARE after becoming eligible for Medicare.

Figure 2.1
Model Fit for Enlisted Personnel

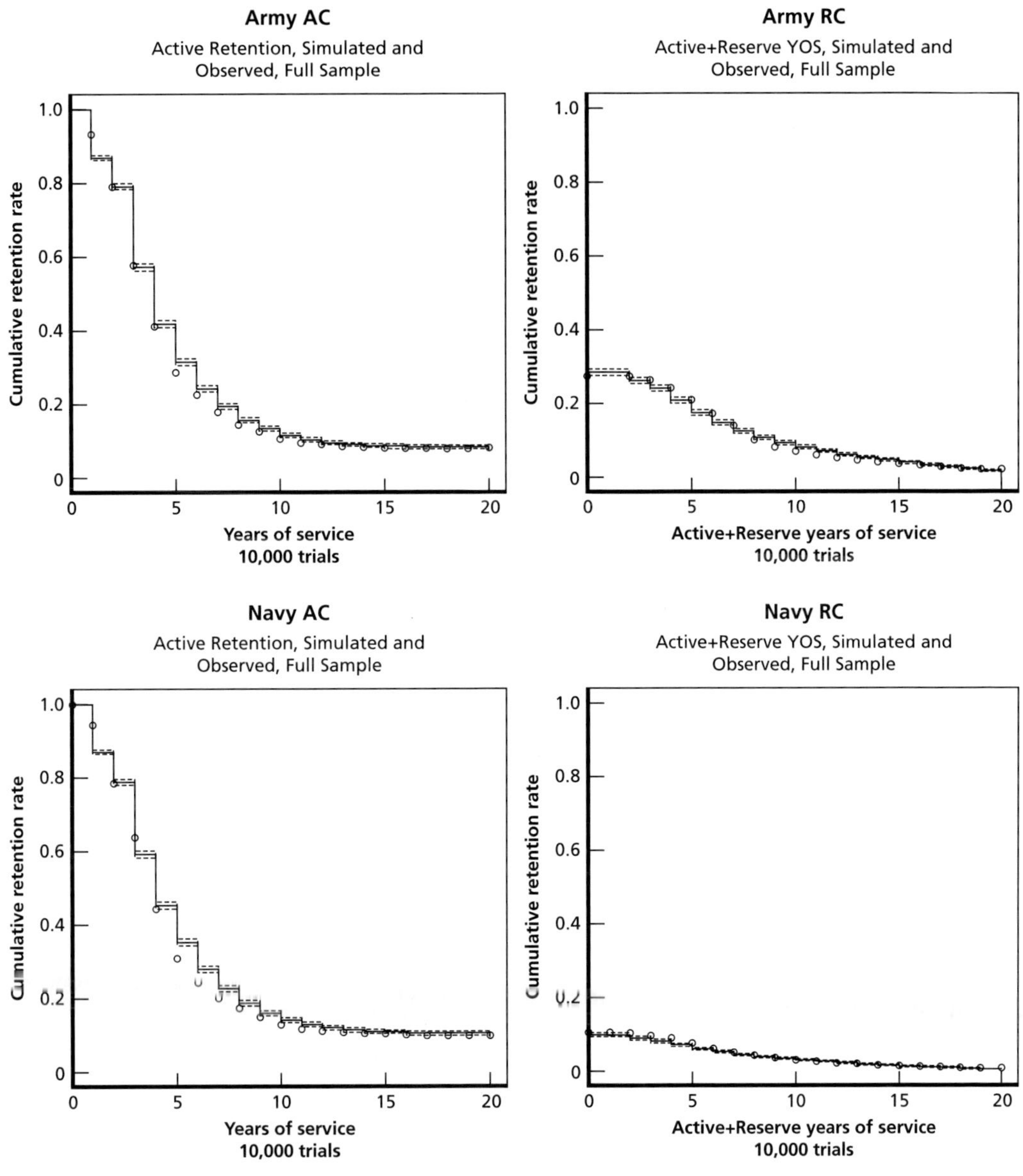

RAND *MG378-2.1a*

NOTES: The circles in each figure represent actual retention and the solid line shows simulated retention. The dotted lines show the error bands.

Figure 2.1—Continued
Model Fit for Enlisted Personnel

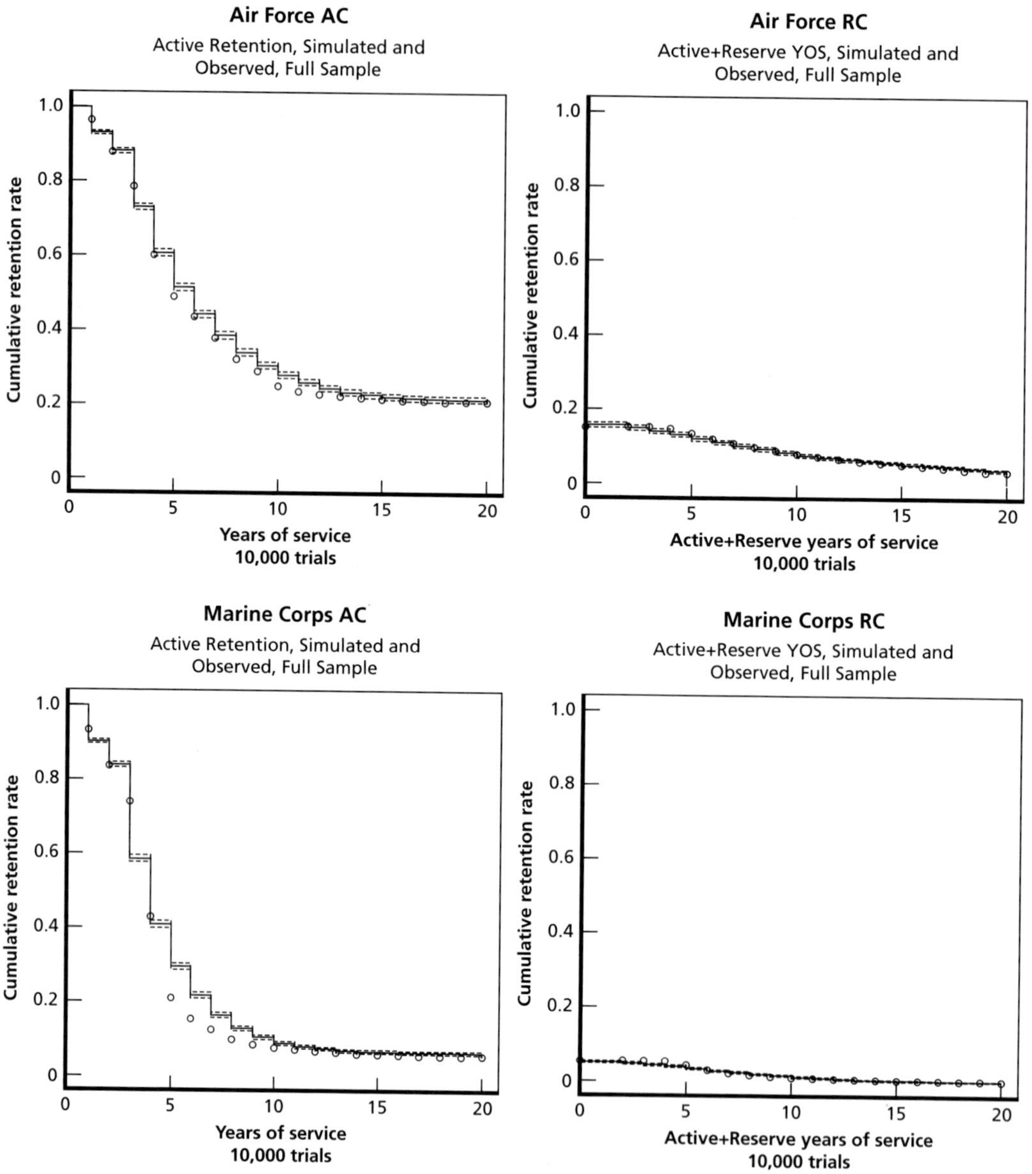

RAND *MG378-2.1b*

NOTES: The circles in each figure represent actual retention and the solid line shows simulated retention. The dotted lines show the error bands.

Figure 2.2
Model Fit for Officers

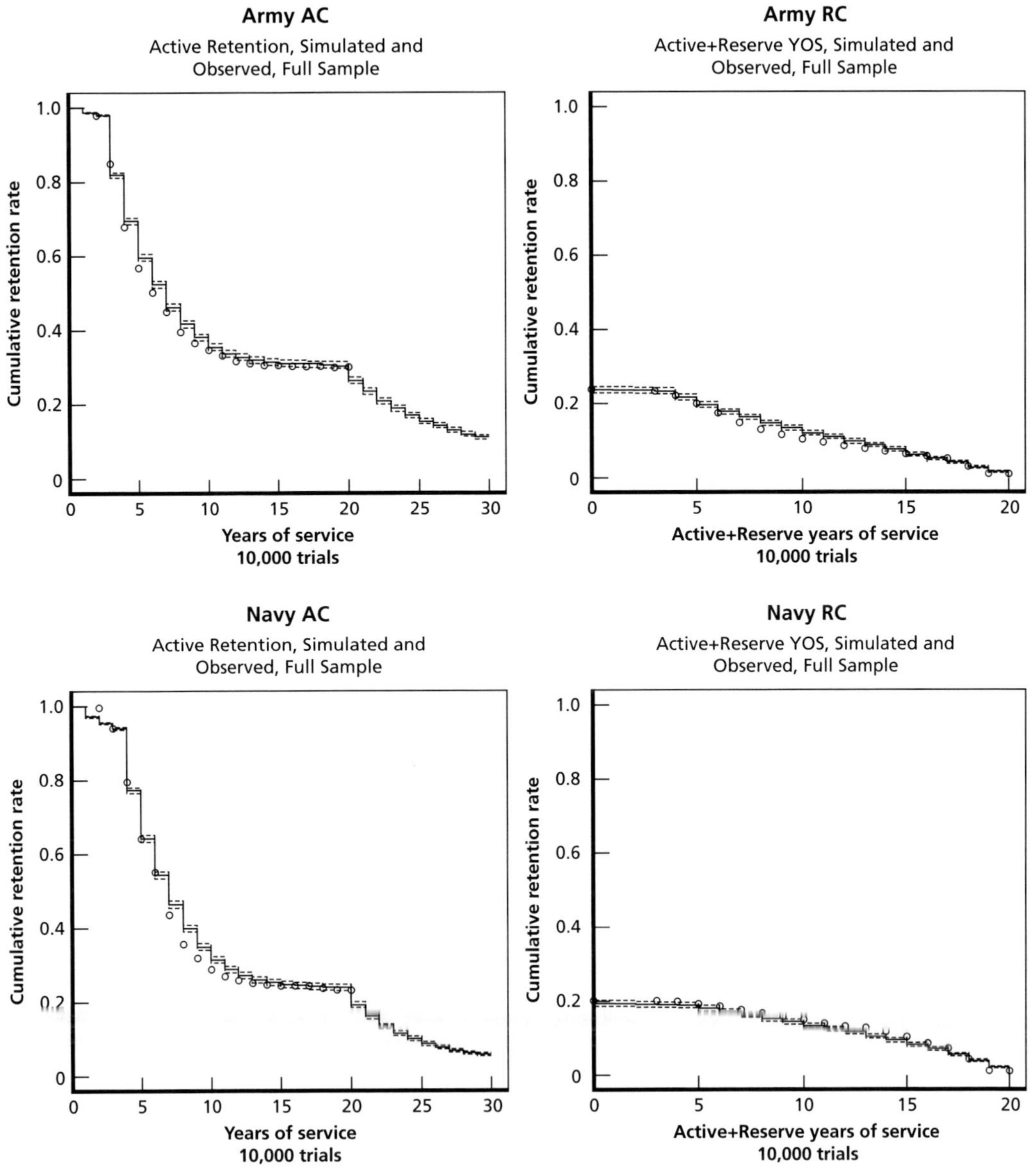

RAND *MG378-2.2a*

NOTES: The circles in each figure represent actual retention and the solid line shows simulated retention. The dotted lines show the error bands.

Figure 2.2—Continued
Model Fit for Officers

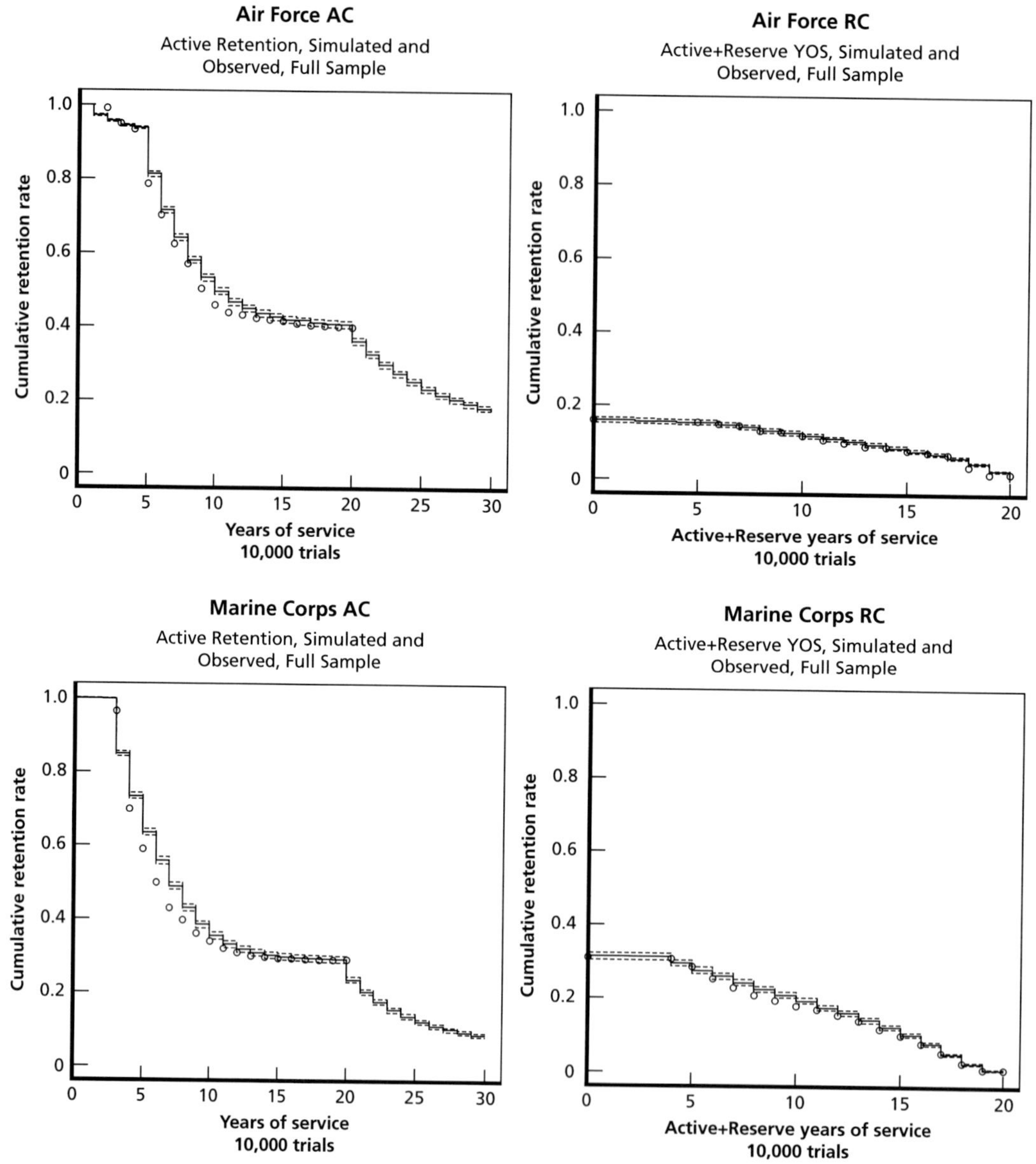

NOTES: The circles in each figure represent actual retention and the solid line shows simulated retention. The dotted lines show the error bands.

CHAPTER THREE

Analysis of Congressional Proposals

The congressional proposals described in Chapter One increase the generosity of reserve retirement benefits by lowering the age of first receipt to 55, allowing receipt of benefits to begin immediately upon leaving the reserves with at least 20 YOS, or lowering the age of first receipt by one year for every 2 YOS beyond 20 to a minimum of age 53. The proposals can be interpreted as a means to increase the compensation of long-serving reservists in view of the increased use of reserve forces. This chapter presents our simulation results of these three proposals. To establish a context for assessing the retention changes caused by the proposals, we begin by simulating a basic pay raise of 1 percent across the board. This affects participation in both the actives and the reserves and allows us to describe how the model works. We then focus on the congressional proposals.

Across-the-Board Pay Raise

Active and reserve components use the same basic pay table, so the 1 percent pay raise benefits both actives and reserves. The pay raise is large enough to affect personnel dynamics in the reserves as well as the actives. Figures 3.1 and 3.2 show the effect on Army AC and RC members caused by the increase in basic pay, relative to the base case. The top panel of each figure shows the number of members and the bottom shows the change in the number when pay changes relative to the base case. The base case is defined as the current basic pay table as of 2007 and the current, High-3 retirement benefit system, as described in Chapter Two. The figures are based on the simulated behavior of 10,000 individuals followed over their careers in the AC enlisted Army and then the Army reserves, if they join; everyone is assumed to start in the active-duty enlisted Army at age 20 and then followed over a 40-year work life to age 60. As discussed in the previous chapter, the simulations are scaled up to the AC force sizes in 2009.

As seen in Figure 3.1, the 1 percent across-the-board increase in pay causes more individuals to stay on active duty. As seen in the top panel, the red line indicating simulated retention under the higher pay policy is just slightly above the blue line for

Figure 3.1
Simulated Army Enlisted AC Retention Due to 1 Percent Pay Increase, Total Retention and Change in Retention

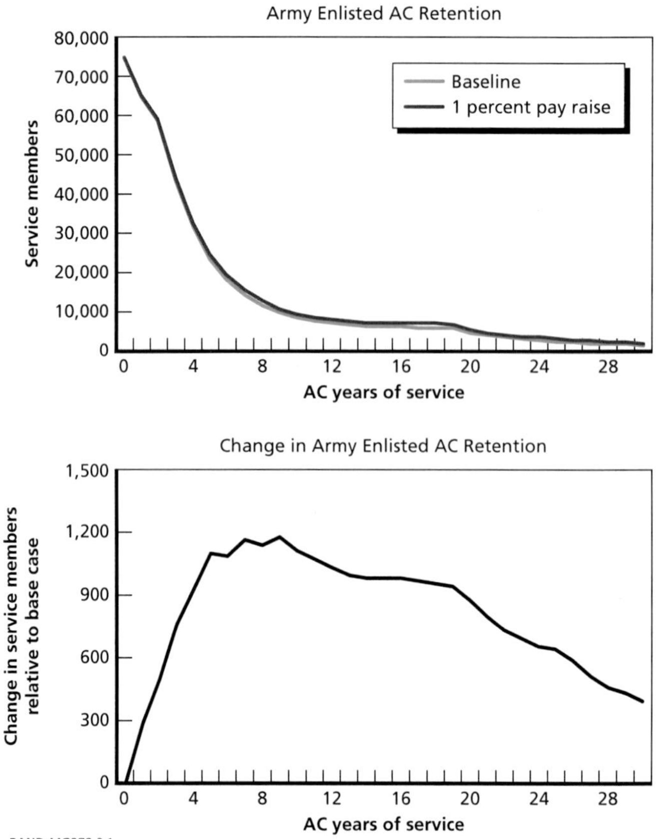

RAND *MG378-3.1*

the base case. The effects are easier to discern by considering the change in retention shown in the bottom panel. AC retention increases by about 1,100 to 1,200 soldiers between 5 and 11 YOS and by about 1,000 until 20 YOS. The increase in basic pay not only increases compensation in each year of service, but also increases the value of the retirement benefit. Nonetheless, the effect on retention is lower after 20 YOS.

Overall, AC Army enlisted force size increases by 5.7 percent, with increases of 2.9 percent at 4 YOS, 9.5 percent at 8 YOS, and 17.5 percent at 20 YOS. These percentage increases are higher than found in cross-section retention studies (see Goldberg, 2001) and can be attributed to the lifecycle nature of our analysis and permanence of the basic pay increase. In the lower panel of Figure 3.1, the increase in retention attained by 5 YOS results not only from more members at 5 YOS choosing to stay, but also from more members staying in the first through fourth year of service. Further, our model captures the fact that a permanent, across-the-board increase in basic pay increases the present discounted value of retirement benefits, and this increase becomes increasingly valuable as the member nears 20 YOS. By comparison, a typical cross-section retention study focuses on behavior at a reenlistment decision point and does not analyze possible retention effects in earlier years that could cumulate as they do in our steady-state analysis. Also, our analysis, by construction, assumes that the basic pay increase is permanent, while members at a reenlistment decision point might believe that a basic pay increase will not represent a permanent increase in military pay relative to civilian pay. For these reasons, the response to an increase in basic pay is likely to be less in a typical retention study than in our simulation.

Figure 3.2 shows the effect of the pay increase on reserve participation (again, total active plus reserve years of service). As the figure indicates, fewer individuals join the reserves than under the base case, although we note that the scale of Figure 3.2 is quite different than that of Figure 3.1. This decrease is a counterpoint to the increase in active-duty retention—and the greater number qualifying for active-duty retirement. The pay increase causes a larger absolute increase in active-duty career earnings and

Figure 3.2
Simulated Army Enlisted RC Participation Due to 1 Percent Pay Increase, Total Participation and Change in Participation

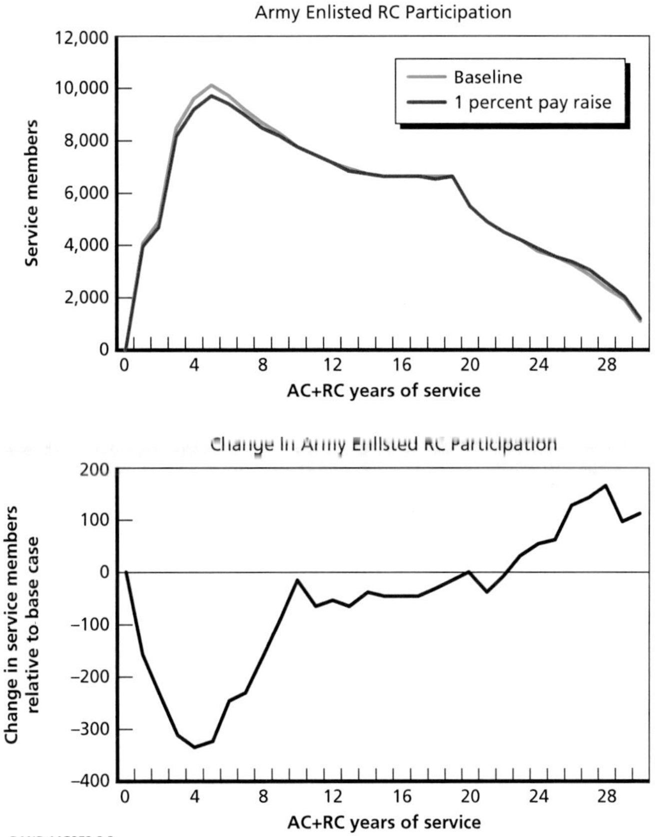

RAND *MG378-3.2*

retirement benefits than in reserve career earnings and retirement benefits, so more personnel choose to remain on active duty and qualify for retirement. Apart from the decrease in reserve affiliation, i.e., in ever joining the reserves, fewer reservists participate before 20 YOS. However, participation increases among those with more than 20 YOS. Nonetheless, overall, Army enlisted RC participation falls and RC force size falls by 1 percent. The pattern of results is quite similar for the other services (not shown).

As expected, the results are consistent with past research showing that an increase in pay also increases active-duty retention (Hansen and Wenger, 2005). However, as mentioned, past research usually focuses on how pay increases affect retention at a particular career point, such as the first or second reenlistment point. Since our model considers the retention response over the career, the results are not directly comparable. Furthermore, no research is available on how RC participation among those with prior AC service is affected by pay, accounting for the effect of a pay increase on AC retention. Our results provide new insights into this question and show that RC retention among this population declines, especially among those who are not qualified for retirement benefits, because fewer AC personnel leave the actives and join the reserves. These results suggest that understanding the responsiveness of RC members with prior active service to changes in their compensation system, such as their retirement benefits, must account for the responsiveness of AC members as well. As we will see in the next subsection, changing the RC retirement system can affect AC retention as well as RC participation and both effects can be important for understanding the effects of compensation changes that affect RC members.

Effects of Force Management

The past congressional proposals would reduce the age at which reservists may begin claiming retirement benefits. This increases the value of reserve retirement benefits in two ways: Benefits are received for more years, and the benefits that are received sooner are discounted for fewer years and so appear larger in present-value terms. We first show the effects of each proposal on AC retention, by service, for enlisted personnel and officers. We then turn to the effects on RC participation.

Effect on Active Retention

The three congressional proposals differ in their generosity. Although more generous than the current reserve retirement benefit system, the age-55 annuity and the sliding-scale annuity are considerably less generous than the immediate annuity. The immediate annuity is the most generous because the benefits are discounted for fewer years and received for more years. The difference in generosity causes different magnitude of effects on active-duty retention. This subsection presents the simulated change in AC

Figure 3.3
Simulated Change in Army Enlisted AC Retention Under Each Congressional Proposal

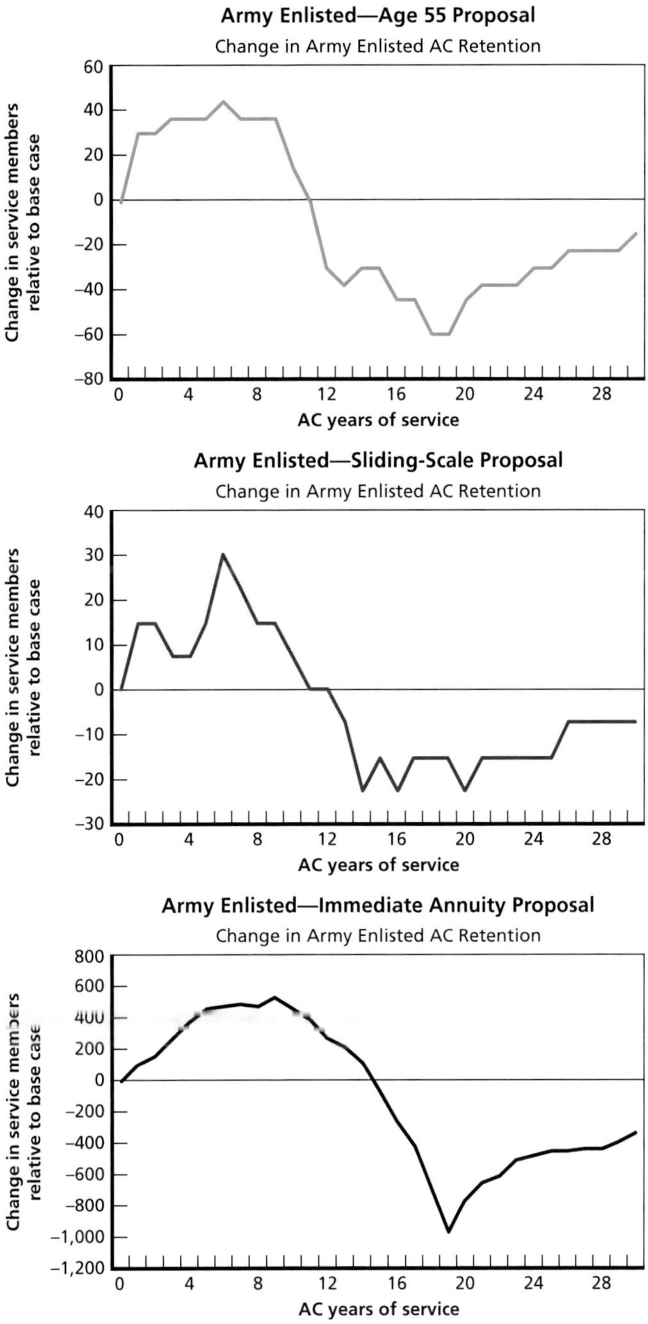

RAND *MG378-3.3*

retention under each proposal for enlisted personnel and officers, by service. Figure 3.3 shows the results for each proposal for Army enlisted RC personnel. We then summarize the results for all three proposals, by service and for enlisted personnel and officers, in Figure 3.4.

As seen in Figures 3.3 and 3.4 for Army enlisted personnel, the effects of the age-55 annuity and the sliding-scale annuity are slight. Active-duty retention increases slightly among those with less than 20 YOS. Figure 3.3 shows the change in AC retention under each proposal separately in different charts for Army enlisted personnel, while Figure 3.4 shows the change under each proposal in a single chart for each service and for enlisted and officers. (It is important to note that the scale of the y-axis differs across charts throughout this chapter, including Figures 3.3 and 3.4; keep this in mind when comparing effects across charts.)

While Figure 3.3 shows some change under the age-55 proposal and the sliding-scale proposal, Figure 3.4 makes it clear that the change is quite small, especially in comparison to the change under the immediate annuity, discussed shortly. Intuitively, the reason for the increase under these proposals is that increasing the generosity of reserve retirement benefits acts indirectly to increase the value of remaining on active duty. If one should choose to leave active duty in a future period, there would be a shorter time to qualify for reserve retirement benefits. Still, these effects are small. In the case of the age-55 proposal, the fraction of Army enlisted personnel who qualify for active-duty retirement changes by less than 0.1 percent, while in the case of the sliding-scale proposal the change is virtually zero.

The immediate reserve annuity causes a larger increase in the number of active-duty members who stay in the midcareer years between the fourth and 12th year of service. Yet now, under the immediate reserve annuity, the draw of reserve benefits is sufficiently strong to induce some service members to leave the actives after year 15, thereby qualifying for reserve retirement benefits. Although their retirement benefits will not be as high as if they had stayed on active duty, those individuals who leave active duty for the reserves anticipate that the reserve annuity will make them better off than staying on active duty. Consistent with this view, we see a decrease of about 1,000 in the number of service members with 20 YOS in the bottom of Figure 3.3. The net effect is that AC Army enlisted force size is nearly unchanged, falling by less than 1 percent.

The pattern of results for officers and for the other services is the same, though the magnitudes differ. For example, the magnitudes of change in terms of number of service members tend to be smaller for officers. Comparing results across services, the changes tend to be somewhat larger for the Army and Navy. Nonetheless, in each case, the magnitude of change under the age-55 and sliding-scale proposals is smaller than the magnitude under the immediate annuity proposal.

In sum, the age-55 and sliding-scale annuity proposals have a small effect on the AC force. The immediate annuity proposal has a larger effect, resulting in a somewhat

Figure 3.4
Simulated Change in AC Retention Under Each Congressional Proposal, by Service, Enlisted and Officers

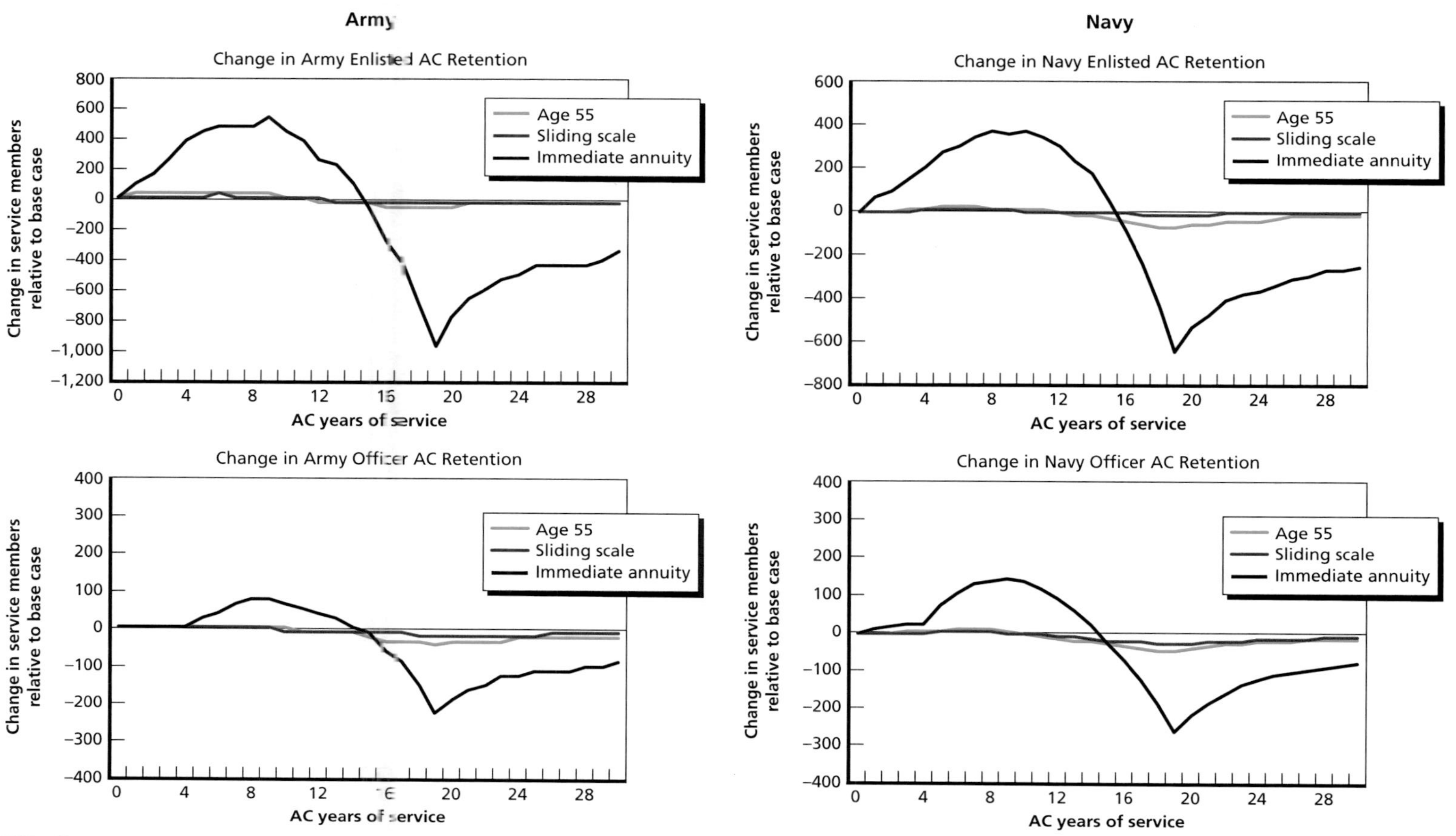

RAND *MG378-3.4a*

Figure 3.4—Continued
Simulated Change in AC Retention Under Each Congressional Proposal, by Service, Enlisted and Officers

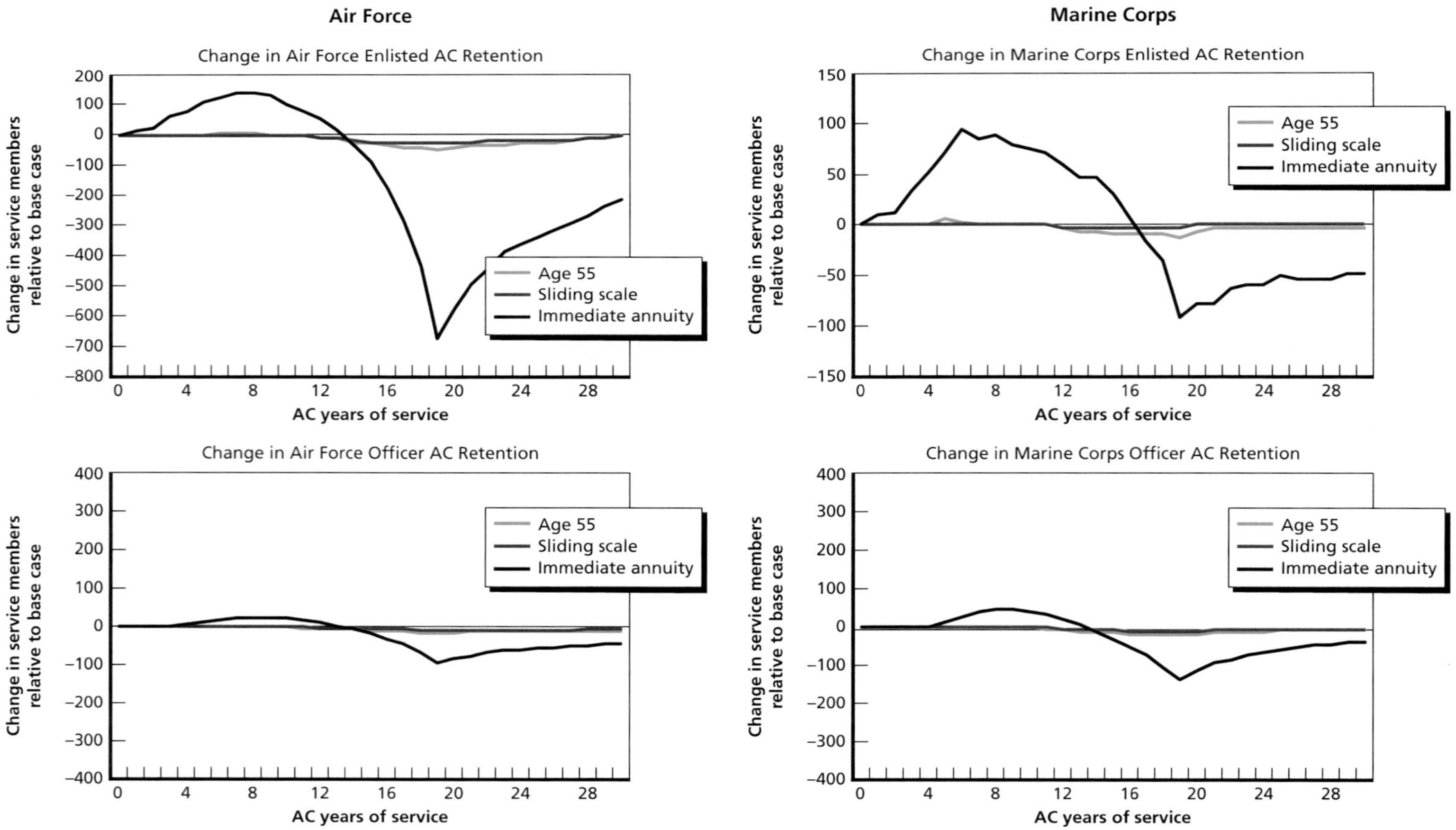

RAND *MG378-3.4b*

Figure 3.5
Simulated Army Enlisted AC Retention, Baseline and Immediate Annuity Proposal

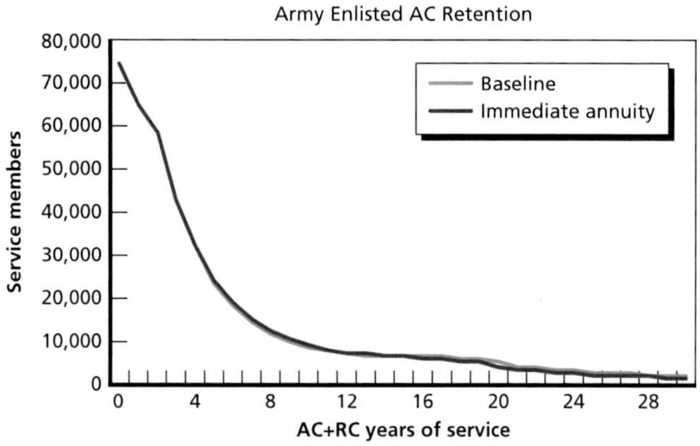

RAND *MG378-3.5*

more junior force and specifically in fewer AC members qualifying for retirement. While the immediate annuity proposal effects are clearly larger, the overall change in the AC force profile is still quite modest. This can be seen in Figure 3.5, where we show the AC retention profile under the baseline versus the immediate annuity proposal for the Army enlisted force. (This is in contrast to Figures 3.3 and 3.4 where we consider the change in retention, not the level of retention). As can be seen, the new AC retention profile is quite similar to the baseline, indicating that the scale of the AC retention changes is still quite modest, even for this proposal. The results for the other services and for officers are qualitatively similar.

Whether even small changes in the AC force would be acceptable to the services is an open question. They may desire to maintain the current baseline force profile and opt to use special and incentive pay to sustain AC retention. Doing so would increase the cost of the AC force. We do not consider such policy changes in our analysis of cost below.

Effect on Participation in the Reserves

Figure 3.6 shows the simulated effects of each proposal on Army enlisted participation in the Army National Guard and Reserve where we use separate charts for each proposal to make clear the pattern of changes relative to the base case. Figure 3.8 shows the effects for each proposal together, by service, for enlisted personnel and officers. These figures show the change in RC participation, given a member with prior AC service joined the RC. (We discuss the purpose of Figure 3.7 shortly.) Again, the scale of the y-axis differs across charts so care is needed in making cross-chart comparisons.

It is also of interest to consider how the proposals affect the flow of AC personnel to the RC. Calculating the flow rate over a military career is challenging because a

Figure 3.6
Simulated Change in Army Enlisted RC Participation Under Each Congressional Proposal

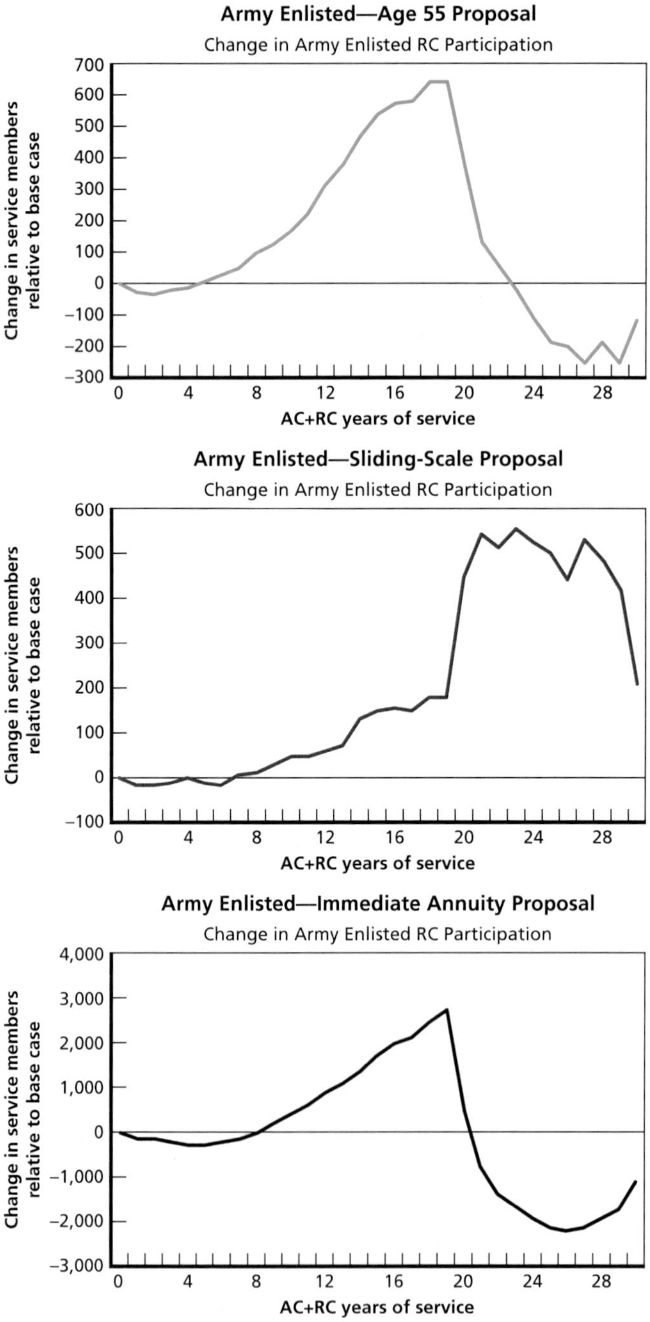

RAND *MG378-3.6*

member may join the RC, leave, and later rejoin the RC. Thus, a simple count of the number of joiners may overstate the flow of people who ever join the RC because some people may join more than once. Another complication is identifying the benchmark by which to compare the flow. Some analysts consider the flow among those who separate from the AC. But, in a lifecycle model such as ours, eventually everyone leaves the AC. We adopt the following convention: We measure the number of people who ever join the RC as a fraction of people who enter the AC. Thus, if 10,000 people join the AC, and 3,800 people with prior AC service ever join the RC, we define the affiliation rate as 38 percent. Table 3.1 shows the affiliation rate under our baseline case and the percentage change in the affiliation rate under the three retirement proposals.

The age-55 proposal has a small effect on prior-service affiliation with the Army enlisted reserves. For example, our simulated results indicate that the affiliation rate among AC entrants increases from 36.2 percent to 36.4 percent, about a 0.6 percent increase. Nonetheless, as seen in Figures 3.6 and 3.8, given affiliation, more reservists with between 5 and 20 YOS in the AC plus the RC stay in the reserves, relative to the base case. Further, more RC members participate after 20 years until 23 YOS, although the increase is not as great as before 20 YOS. Thus, being allowed to claim RC benefits five years earlier than age 60 induces some RC members to participate more even after qualifying for RC retirement benefits at 20 YOS.

We can view the results somewhat differently if we consider the change in RC participation by age, or by years since starting with the AC. In our model, everyone begins their AC service at age 20. Thus, years since joining the AC equals years since age 20. These results are shown in Figure 3.7 for the Army enlisted personnel. The x-axis is age (less 20 years) or years since starting active duty. The graph shows the

Figure 3.7
Simulated Change in Army Enlisted RC Participation Under Congressional Proposals, by Years Since Entering AC

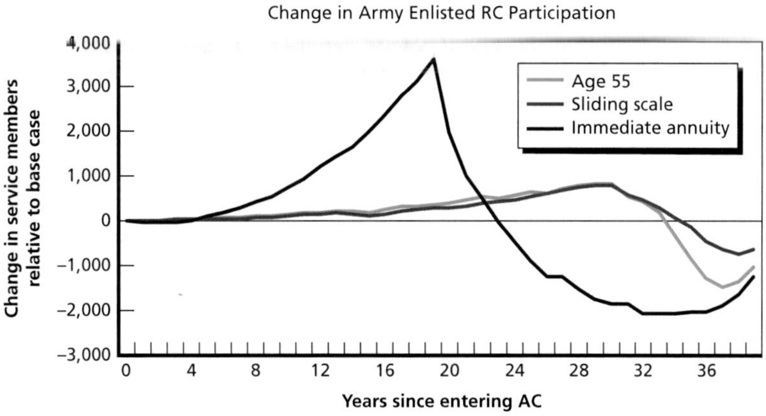

RAND *MG378-3.7*

change in RC participation. This is in contrast to Figures 3.6 and 3.8, where the x-axis is active plus reserve years of service; note years since joining the AC. We find that the age-55 proposal increases RC participation particularly between years 12 and 33, corresponding to ages 32 to 53. Upon reaching the end of age 54 (or the beginning of age 55, corresponding to year 35), RC participation declines. This is as expected because the age-55 proposal allows members to begin claiming benefits at age 55. We find that when members reach this age, they leave to claim benefits, but their participation increases prior to age 55 because the ability to claim benefits at an earlier age increases the value of the benefit to them, so the value of staying in the RC increases. Thus, roughly speaking, more of the older reservists now qualify for retirement benefits at age 55 and terminate their reserve careers to claim them.

Figure 3.8 shows a similar pattern for the age-55 proposal for the other services and for officers. These charts show the change in participation by years of AC service plus RC service rather than by age (or by years since start of AC). The pattern of results based on age (not shown) is also quite similar to the Army enlisted for the other services and for officers.

The results for the sliding-scale annuity have similarities with those for the age-55 proposal, although the effects of the sliding-scale annuity are even smaller and less prominent. Prior-service affiliation with the reserves is unchanged, as shown in Table 3.1. There is some increase in reserve participation among those whose years of AC plus RC service is less than 20, but much of the increase that does occur in participation occurs after 20 YOS. The sliding-scale proposal rewards those who stay longer by reducing the age at which the RC retirement benefit can be claimed. This induces individuals who have qualified for retirement to stay longer. However, as seen in

Table 3.1
Baseline and Percentage Change in Selected Reserve Affiliation Rate of Personnel with Prior Active-Duty Service Under Congressional Proposals

		Percent Change from Baseline		
Enlisted	**Baseline**	**Age-55 Annuity**	**Sliding-Scale Annuity**	**Immediate Annuity**
Army	36.23	0.55	0.06	5.13
Navy	14.28	2.73	0.49	19.19
Marine Corps	7.06	1.13	0.28	7.51
Air Force	17.96	1.95	0.67	17.87
Officer				
Army	27.8	2.95	1.15	15.50
Navy	26.2	6.07	2.29	34.73
Marine Corps	36.7	4.90	1.91	31.34
Air Force	20.5	5.12	2.49	28.88

NOTE: RC affiliation rate in the table is defined as the percentage of AC entrants who ever join the RC.
SOURCE: Authors' simulations.

Figure 3.8
Simulated Change in RC Participation Under Each Congressional Proposal, by Service, Enlisted and Officers

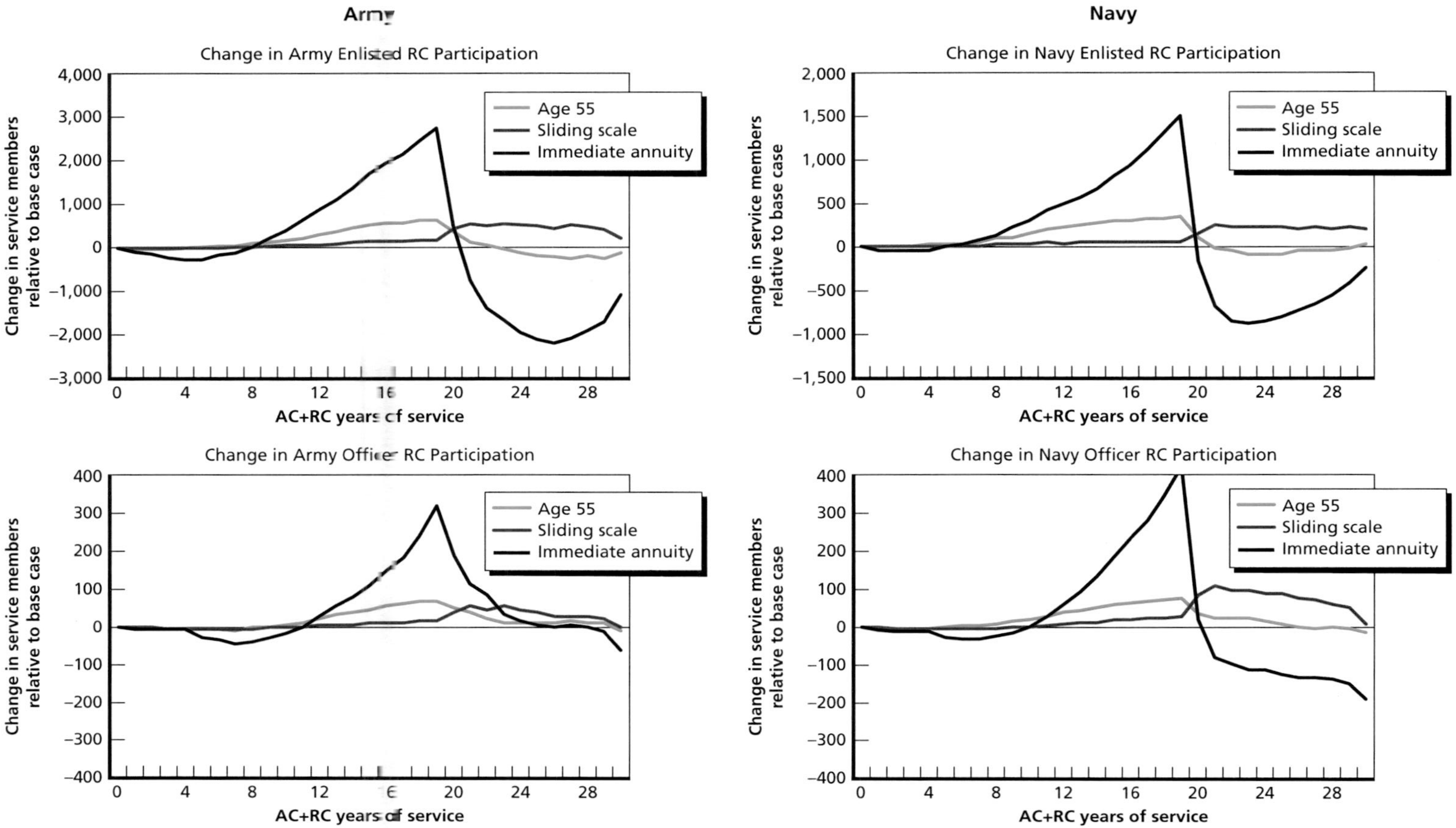

RAND *MG378-3.8a*

Figure 3.8—Continued
Simulated Change in RC Participation Under Each Congressional Proposal, by Service, Enlisted and Officers

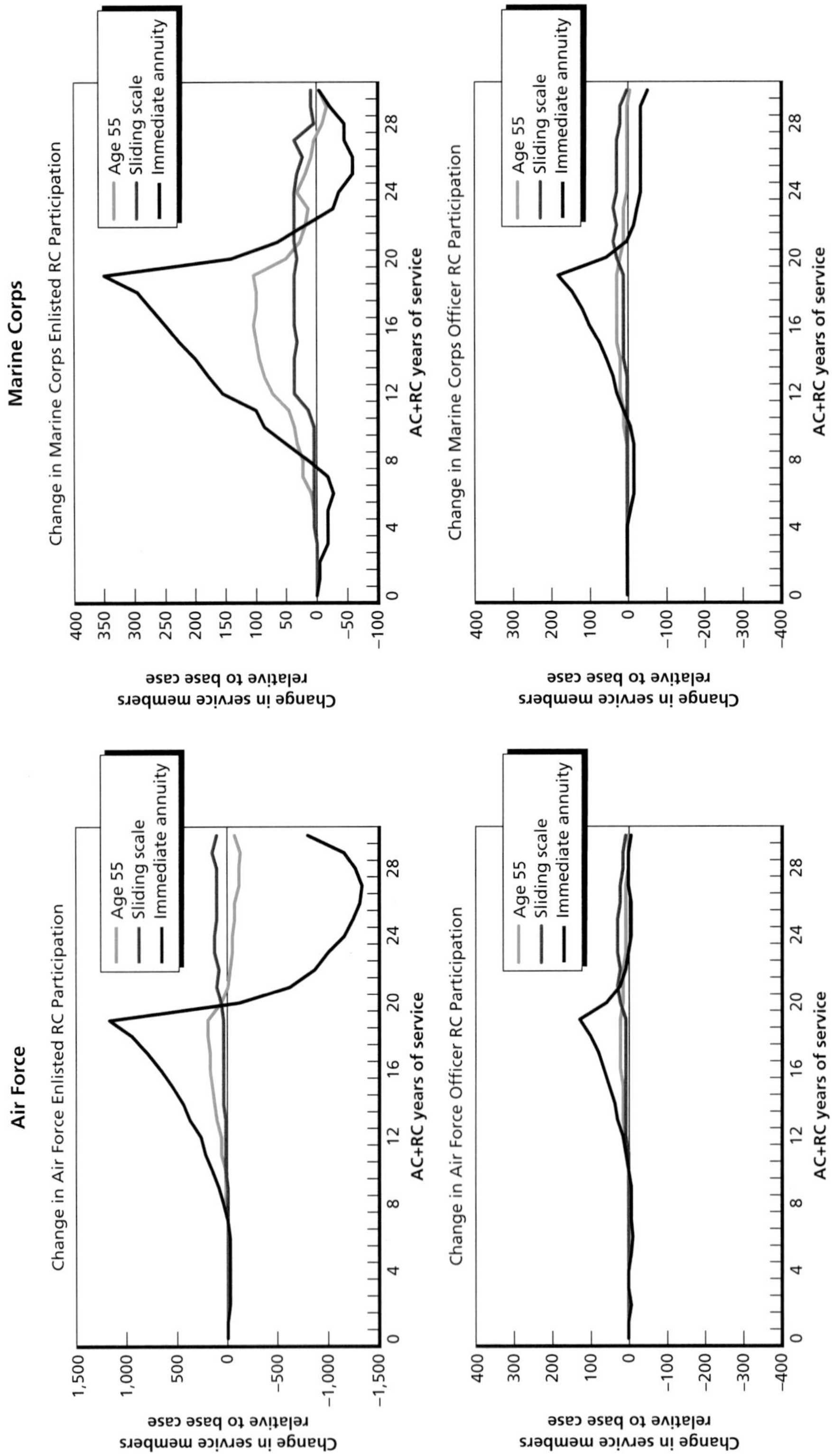

Figure 3.7 for the Army, which shows the change in participation by age, eventually members are able to claim benefits at an age earlier than age 60, and once they reach those ages, they are more likely to leave. In Figure 3.7, we see that beginning at age 56, participation declines. Since this proposal allows retirement one year earlier for every two years served beyond 20, age 56 corresponds to someone who has 28 YOS (28 – 20 = 8) and has earned the right to retire 4 years earlier (60 – 4). Again, because individuals can now claim benefits, they do so and leave.

It is interesting to contrast the results under the sliding scale with the age-55 proposal. The age-55 proposal induces greater participation particularly among those with less than 20 YOS, while the sliding-scale proposal induces greater participation particularly among those with at least 20 years, although the magnitude is not large under either case. Under the sliding-scale proposal, to gain a one-year decrease in the age of first receipt of retirement benefits, the individual must have two more years of service, either in the actives or in the reserves. Thus, the gain from reducing the age of first receipt of benefits by a year always comes with an incremental obligation to serve two more years. In contrast, the age-55 annuity has no such incremental obligation. The sliding-scale proposal therefore "pulls" those with 20 YOS toward a longer career while the age-55 proposal "pulls" those with less than 20 YOS to stay and claim what is now a more valuable benefit (since it can be claimed at a younger age) but it also pushes people out once they reach age 55. Again, while these patterns are clearly discernible, the magnitudes of these effects are small, especially compared with the immediate annuity case. For example, the increase in RC participation after 20 YOS in the sliding-scale case is only about 400 members per year for the Army enlisted force.

The effects of the immediate annuity proposal are the largest of the three proposals. Prior-service accessions of Army members to the Army National Guard or Army Reserve increases by about 5 percent (Table 3.1). The percent increases in affiliation are even greater for the other services and for officers. RC participation also increases among those with less than 20 YOS, and the increase is greater as RC members near 20 YOS, as expected. The ability to claim an immediate annuity upon reaching 20 YOS increases the value of the RC retirement benefit and produces a strong draw among those with less than 20 YOS who increase their participation, especially those near 20 YOS.

Under the immediate annuity, retention falls among senior personnel who can begin receiving the reserve retirement benefits as soon as they leave. In Figures 3.6 and 3.8, we observe a substantial drop in RC participation after 20 YOS when they qualify for immediate retirement benefits. In the case of Army enlisted personnel, the change in participation is actually negative. As seen in Figure 3.7, which shows Army enlisted RC participation by years since start of AC service (or by age minus 20), participation drops off at year 20 and becomes negative at year 24 (or age 44). This suggests that in many cases, they have enough reserve service by year 24, given their active-duty service, to qualify for the immediate annuity.

The larger effects of the immediate annuity proposal on RC participation is apparent when we consider the RC career in terms of years of AC service plus RC service (as in Figure 3.6) and when we consider it in terms of years since start of active duty (as in Figure 3.7). Yet another way to see the effect is to consider the RC participation profile under the baseline versus under the immediate annuity proposal. This is in contrast to Figures 3.6 to 3.8 where we show the change in the profiles, rather than the levels. Figure 3.9 shows the level of participation for the Army RC enlisted force under the baseline and under the immediate annuity proposal. The figure shows the large increase in pre-20 YOS participation, followed by a large drop-off once members qualify for immediate retirement benefits. In contrast, the changes under the sliding-scale and age-55 proposals are much more modest.

The effects for the other services are qualitatively similar to those for the Army, shown in Figures 3.6 through 3.8. The immediate annuity proposal has the largest impacts relative to the other proposals; it increases participation the most among junior reservists and decreases it the most among senior reservists.

Effect on Reserve Retirement Rates and Reserve Force Size

Table 3.2 contains additional measures of merit for the proposals, namely the change in the number of RC personnel in each service with prior AC service and the change in the percentage of personnel who ever join the RC that reach RC retirement eligibility of 20 YOS. The table shows these percentages in the baseline, by service and for enlisted personnel and officers, and then the percentage change relative to the baseline.

Under the age-55 proposal, the expected number eligible for retirement goes up because more reservists are retained through the mid- and late-career, as shown in the top panel of Table 3.2. Among Army enlisted personnel who ever join the RC, the

Figure 3.9
Simulated Army Enlisted RC Participation, Baseline and Immediate Annuity Proposal

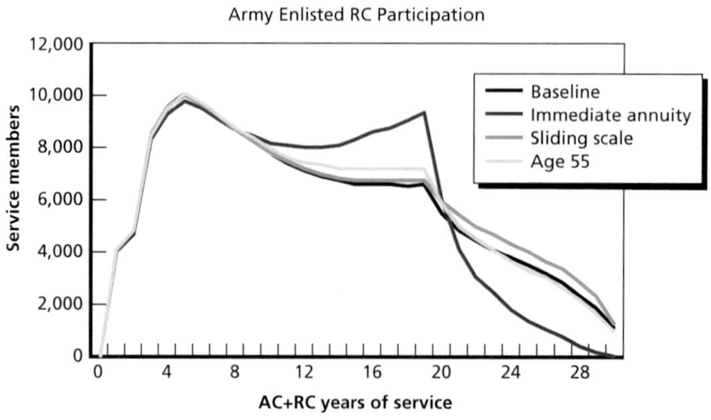

RAND *MG378-3.9*

Table 3.2
Baseline and Percentage Changes in Reservists Ever Joining Who Qualify for Retirement Benefits and in RC Force Size Due to Congressional Proposals

		Percent Change		
Measure	**Baseline**	**Age-55 Annuity**	**Sliding-Scale Annuity**	**Immediate Annuity**
	(Percent ever joining RC that reach reserve retirement benefits)			
Enlisted				
Army	24.29	9.62	2.73	35.43
Navy	37.18	12.56	2.45	42.84
Marine Corps	19.55	22.52	8.03	69.18
Air Force	45.55	6.12	1.28	26.94
Officer				
Army	59.06	4.41	0.97	15.42
Navy	88.00	2.06	0.42	6.42
Marine Corps	75.83	3.34	0.76	11.84
Air Force	81.00	2.48	0.83	8.36
	(Size of RC force with prior AC service)			
Enlisted				
Army	178,895	2.21	3.54	-1.10
Navy	43,466	5.94	6.55	3.30
Marine Corps	12,284	8.95	4.96	16.30
Air Force	54,827	1.38	2.91	-9.54
Officer				
Army	18,617	3.49	2.82	7.43
Navy	17,092	4.13	5.86	2.24
Marine Corps	7,093	4.12	4.77	6.32
Air Force	6,023	4.09	5.12	9.20

SOURCE: Authors' simulations.

number that reach retirement eligibility increases by 9.62 percent; for Army officers, it increases by 4.41 percent. We see similar increases for the other services. The decline in retention due to the increase in retirements at age 55 is not enough to offset the increase in retention among mid- and late-career personnel, resulting in a modest increase in RC force size overall. For example, among Army enlisted personnel, the RC force increases by 2.21 percent. Similar general increases occur for the other services, on the order of around 2 to 4 percent.

The results for the sliding-scale annuity proposal are quite similar, and the changes are even more modest. The percentage of those who ever join the RC increases slightly, as does the RC force among those with prior service.

The immediate annuity causes a large increase in the number of reservists eligible for retirement benefits. The increases are on the order of 27 to 69 percent for enlisted personnel and 6 to 15 percent for officers. These large increases are accompanied by smaller changes in the RC force size. Although more RC members are induced to stay until 20 YOS, more of them also leave upon reaching that milestone. In some cases the net effect is positive and in others it is negative. In the case of Army enlisted personnel, the effects are just offsetting, resulting in a slight decline in RC force size. In contrast for the Marine Corps enlisted force, the net effect is positive, increasing by more than 16 percent.

It is possible that the services will not desire the predicted increase in the size of RC members with prior AC service. They might choose to respond to the increase in participation by adjusting personnel policies. For example, they might restrict recruiting quotas and limit the number of new entrants. Our model shows the predicted change in participation as a result of the changes in the returns to the RC but does not incorporate how the services might respond in terms of changing personnel policies.

Effects on Cost

We next turn to the change in RC costs as a result of the congressional proposals (see Table 3.3). We describe our methodology for computing cost in Chapter Two and in Appendix B. We show cost estimates for the baseline, by service, for RC personnel with prior AC service and we show the change in cost under each proposal relative to the baseline.[1] Total costs include two main components, the RC retirement cost and the cost of RC current compensation, specifically basic pay and any housing and subsistence allowance RC members receive. Under the baseline, RC personnel costs for members with prior AC service are $2.86 billion, with the majority of costs attributable to current compensation. RC retirement costs for those with prior AC service are only $0.29 billion. The table also shows the change in RC cost per member. We focus on cost per member as a way of controlling for cost changes that come from changes in RC force size. That is, the table shows the percentage change in cost, apart from cost changes resulting from the changes in RC force size shown in Table 3.2. Thus, changes in RC costs occur because of changes in the experience mix of RC members.

The congressional proposals increase reserve retirement costs per member substantially in percentage terms. The largest change is for the immediate annuity proposal, where RC retirement costs per member more than triple. However, in our model, reserve costs for RC members with prior AC service are only a small fraction of total retirement costs, on the order of about 2 percent of total retirement costs per member. Furthermore, RC retirement costs are a small fraction of total costs for RC members

---

1 Our model also computes AC cost as well, but these costs are not shown in the table.

Table 3.3
RC Baseline Cost and Changes in Cost per Member: Enlisted Personnel and Officers (2007 billions of $)

	Current Cost	Retirement Cost	Total Cost
Baseline cost			
Army	1.41	0.13	1.54
Navy	0.52	0.08	0.61
Air Force	0.47	0.05	0.52
Marine Corps	0.17	0.02	0.19
Total	2.57	0.29	2.86
(Percentage change in cost per member under policy alternatives)			
Age-55 Proposal			
Army	0.10	40.69	3.54
Navy	-0.46	40.67	5.24
Air Force	-0.07	50.54	4.98
Marines	-0.35	46.36	4.40
Change (percentage of baseline)	0.04	43.40	4.38
Sliding-Scale Proposal			
Army	0.92	26.61	3.10
Navy	1.26	25.69	4.65
Air Force	0.79	33.96	4.10
Marines	0.97	31.02	4.03
Change (percentage of baseline)	1.05	28.38	3.79
Immediate Annuity Proposal			
Army	-1.43	220.31	17.38
Navy	-3.16	247.33	31.57
Air Force	-3.37	327.83	29.67
Marine Corps	-1.35	302.61	29.56
Change (percentage of baseline)	-1.93	254.51	23.75

with prior service. Consequently, the large percentage change in RC retirement costs has a small overall impact on the total RC cost.

The proposals have a small effect on current compensation costs per member. In many cases, current costs decrease under the proposals. Overall RC current costs per member are virtually unchanged by the age-55 proposal because an increase for the Army is offset by a decrease for the other services. Current costs per member increase by 1 percent under the sliding-scale proposal and decrease by about 2 percent under the immediate annuity proposal. Under the immediate annuity proposal, RC participation increases among midcareer personnel with less than 20 YOS, increasing current costs, but this is offset by the drop in participation among those with more than 20 YOS. The net effect is a slight drop in current costs per member.

Table 3.3 also shows the change in total RC costs per member. The age-55 and sliding-scale proposals result in an increase in cost per member on the order of about 4 percent. The immediate annuity generates a substantial increase in cost per member, nearly 25 percent. The increase in total cost per member occurs for all three proposals because of the increase in retirement costs—a result of more RC members qualifying for RC retirement benefits under these proposals and the greater generosity of the proposals relative to baseline. In the case of the age-55 and sliding-scale proposals, the increase is small, but in the case of the immediate annuity, the percentage of members who join the RC and eventually reach retirement eligibility increases dramatically, as shown in Table 3.2.

The main result from Table 3.3 is that the proposals are not cost-effective methods of sustaining the overall size of the RC prior-service force. By considering cost per member, we hold RC force size constant, and our finding that cost-per-member increases under each proposal indicates that these proposals are more-expensive ways of maintaining the same RC force size. In the case of the immediate annuity, the cost per member increases substantially. On the other hand, the force profile changes substantially under the immediate annuity proposal, resulting in a force with a higher proportion of mid-career personnel reaching RC retirement eligibility and fewer senior personnel. To the extent the services desire such a force profile, then the immediate-annuity proposal may be desirable. Thus, each service must evaluate these proposals in light of its manpower requirements and the effects of these proposals on cost, force size, and experience mix, and on the flow of AC personnel to the RC.

Summary

The greater usage of the RC as an operational force could threaten to reduce retention and make recruiting more difficult. Would an increase in the generosity of reserve retirement benefits as proposed be a helpful response?

Results of our policy simulations indicate that the age-55 and sliding-scale proposals have a small effect: They increase retirement costs, have little effect on active-duty retention, and little effect on joining the reserves after active duty. The age-55 proposal causes a small increase in midcareer reserve retention and more members qualify for reserve retirement, but this is offset by a decrease in the number of reservists choosing the longest reserve careers. Overall, RC force size increases a bit under the age-55 proposal and it increases even less under the sliding-scale age proposal. With an increase in cost per member and a small change in RC participation, these proposals do not seem cost-effective.

The immediate annuity proposal is more generous than the age-55 and the sliding-scale proposals and therefore more costly per member. Our simulations indicate that the immediate annuity proposal increases AC retention among midcareer members, with the largest effect between the fourth year and the 12th, but the effect is lower after the 15th year of service as members leave to qualify for reserve retirement. Individuals participate more in the active component and then join the reserves. Reserve affiliation increases substantially under this proposal. Of those who ever join the reserves, reserve participation increases primarily in the range between 8 and 20 YOS and declines thereafter. The increase in midcareer reserve participation and the subsequent decline in senior reserve participation reflect a behavioral response to the opportunity to receive reserve retirement benefits immediately rather than waiting until age 60. RC force size decreases among Army and Air Force enlisted personnel and increases among officers and Navy and Marine Corps enlisted. The immediate annuity proposal does not appear to be cost-effective. RC total costs per member increase by almost 25 percent, resulting from a large increase in the percentage of RC members reaching retirement eligibility and the greater generosity of the retirement benefit.

The final analysis of these proposals requires an assessment by each service of the gain or loss to the service from the change in AC retention and the change in RC affiliation, participation, and cost. For the immediate annuity, each service must assess how they evaluate the loss of active duty mid- and senior-career personnel, the increase in individuals entering the reserves, the increase in midcareer reserve participation, and the decrease in long reserve careers. The midcareer retention might be an advantage to the service if the current retirement benefit system were thought to result in too few service members staying on active duty until they qualified for retirement benefits, and if the immediate annuity offered by the reserves caused just the right number of active-duty service members to leave after 20 YOS. However, the services do not necessarily believe there are too few active-duty members in the mid- and senior career, and even if they held that belief, we know of no studies estimating how many service members should be in these ranges.

More broadly, if the core issues motivating reserve retirement reform include the experience mix of the actives, the flow of experienced active-duty personnel to the reserves, and reserve retention, then we believe it would be worthwhile to consider a range of compensation changes in place of, or possibly in addition to, the retirement proposals. For example, if AC retention in the mid- and senior career were thought to be too low, the services might introduce special and incentive pay targeted at skills in undersupply; if prior-service recruiting into the reserves were a concern, a reserve affiliation bonus could be introduced; and if reserve retention were low because of the high pace of deployments, an increase in deployment pay could be implemented.

CHAPTER FOUR

Reserve Retirement Reform: Discussion of Broader Issues

While the dynamic retention model focuses on the recruiting and retention implications of past congressional proposals, other issues play into the debate over reserve retirement reform, such as equity with the active-duty retirement system. Furthermore, all three of the congressional proposals focused reform on the same area, changing the annuity age of reserve retirement, but there might be alternative proposals that could meet a broader set of reserve compensation goals. In this chapter, we discuss reserve retirement reform in relation to five such goals:[1]

1. Enhancing equity
2. Recognizing more-frequent and longer deployments
3. Ensuring an adequate supply of high-quality reserve personnel with requisite skills and experience
4. Improving flexibility to manage reserve personnel
5. Ensuring a cost-effective military compensation system.

These goals are in the spirit of the general principles underlying military compensation as developed in the 6th and 7th Quadrennial Review of Military Compensation, and they focus on specific concerns, such as equity and deployment, that are part of the policy discussion about reserve retirement reform.

Equity

Few things undermine morale more in any organization than a sense among workers that they are compensated unfairly. The need to convey a sense of fairness among its members is reflected in the relatively rigid schedule by which the military determines basic pay and allowances and determines promotions. As stated in the fifth edition of the *Military Compensation Background Papers*, the principle of equity encompasses two concepts: comparability and competitiveness [DoD, 1996]. Comparability implies

[1] This chapter draws from Asch, Hosek, and Loughran (2006).

that individuals within the uniformed services receive equal pay for equal work. Competitiveness implies that military members receive pay that is competitive with civilian opportunities. But this is not an official definition of equity, and other aspects of equity also deserve attention, such as the distinction between equality of opportunity and equality of outcome. Equality of opportunity can go hand-in-hand with inequality of outcome, provided that individuals are given, and believe they have, equal opportunity and that the outcomes are fair. In the military, some disparity in outcomes is commonplace, such as the difference in the speed of promotion among individuals in a specialty or the payment of special or incentive pay in certain circumstances (e.g., selective reenlistment bonuses).

To help understand equity with regard to reserve retirement, we begin with a brief discussion of three differences between reserve and active-duty retirement: the age of pension receipt, the calculation of pro rata years of service, and the calculation of basic pay for purposes of retirement.

Age of Pension Receipt

As previously discussed, active-duty personnel qualify for retirement benefits when they complete 20 years of active-duty service, and they receive those benefits as soon as they retire from service. A person entering active duty at age 20 and retiring after 22 YOS will receive retirement benefits at age 42.

Reserve personnel qualify for retirement benefits when they complete 20 years of creditable service, and they receive retirement benefits at age 60. Creditable service includes each year of active service, if any, and each year of reserve service in which the individual earned at least 50 points. A reservist receives 15 points for affiliation with the Ready Reserve and a point for each training drill (typically four drills on one weekend each month), each day of active training (typically 14 days each summer), each day of duty when activated, and each day of various other activities such as participation in a funeral color guard. Most selected reservists have no trouble accumulating 50 points over a full year. A person entering active duty at age 20, separating after eight years, immediately joining the Selected Reserves, and participating 14 years continuously will have 22 years of creditable service at age 42. But this reservist will not receive retirement benefits until age 60.

The rules for determining the age of pension receipt seem to favor active-duty personnel. Because many active-duty members retire in their mid-forties, they can receive retirement benefits for about 15 years more than retired reservists. Earlier receipt increases the present discounted value of lifetime retirement pay for AC members relative to that of reservists because there are more years of benefits, and benefits that are received sooner are discounted less.

Pro Rata Years of Service

The reserve retirement benefit calculation is based on pro rata years of service, which equal the number of active years plus the total number of points accumulated in the reserves divided by 360. Someone who serves 10 years in the actives and 10 years in the reserves, earning 720 reserve points (an average of 72 points per year), has a total of 12 pro rata YOS (10 + 720/360). The reservist's retirement benefits would then be based on 12 YOS, while an active's retirement benefits would be based on 20 years. This difference, taken together with the later age of pension receipt for reservists, may have added to the perception of unfairness concerning reserve retirement. Pro rata years of service already adjusts for the fact that reserves are not on year-round duty, thus, the argument goes, reservists should be able to receive retirement benefits immediately upon separating from the reserves with at least 20 creditable years.

Contrary to this position, the difference in how reservists earn years of creditable service could be seen as unfair to active members. Non-deploying reservists typically earn about 70 retirement points per year, and 15 of those points, or about one-fifth, are earned just for being affiliated with the reserves. Reservists receive double points for each day of drilling. For active members, a day is a day in terms of points and there are no points for affiliation with an active component.

Basic Pay

The calculation of basic pay for retirement purposes differs for RC and AC members. Basic pay for the purposes of AC retirement is High-3 pay. Thus, if an AC member retires in January 2012, his basic pay for the purposes of retirement will equal his average basic pay between January 2009 and December 2011.[2] Basic pay for reservists who enter the retired reserve upon separating from the Ready Reserve is calculated based on the basic pay in effect for the 36 months preceding age 60.

The calculation of basic pay is to the reservist's advantage for two reasons. First, between the time of a reservist's separation from the reserves and age 60, basic pay might increase faster than the rate of inflation. In the past decade, this has been the case (Hosek, Asch, and Mattock, 2012). Second, reservists in the retired reserve continue to accumulate longevity for the purposes of calculating basic pay. So, for example, a reservist separating from the Ready Reserve as an E-7 with 20 creditable YOS will receive retirement pay (at age 60) based on the basic pay in effect for an E-7 with 26 or more YOS providing he remains in the retired reserve. Together, these differences can translate into noticeable differences in basic pay. For the E-7 just mentioned, FY 2009 basic pay is $3,951 for 20 YOS and $4,521 for 26 or more YOS, an increase of more than $570 a month, or 14.4 percent. This implies the same percentage increase in the present discounted value of retirement benefits.

---

[2] This assumes the individual does not have other months in which his basic pay exceeded the last 36 months of basic pay.

Summarizing the differences, reserve retirement benefits begin at age 60, are based on pro rata years of service, and are calculated using basic pay from the table in effect at age 60 and including longevity increases provided the reservist is in the retired reserve. The first two factors decrease the present discounted value of RC retirement benefits relative to AC retirement benefits while the third factor increases them. Although the third factor increases reserve benefits, the first two factors—later receipt of benefits and fewer (pro rata) years of service—typically mean that RC retirement benefits are less than AC retirement benefits.

Discussion

Granted that RC retirement benefits are less than those of the AC, is this result necessarily unfair? On the criteria of comparability—equal pay for equal work—the answer depends on whether one believes reservists and active-duty members truly perform equal work. Although reservists are now called upon more than ever to perform the same duties and take the same risks as active-duty members in a military operation, reserve and active-duty professions differ in a number of ways.

Reservists can have a civilian career while employed in the reserves. This allows them to develop civilian and firm-specific skills and knowledge contributing to their civilian earning potential relative to that of active-duty members. In addition, many reservists qualify for private pension benefits through their civilian employers. Although we do not have specific figures for reservists, among full-time, full-year wage and salary workers ages 21–64, 66 percent of workers with some college report working for an employer offering a retirement plan, and given that a plan is offered, about 85 percent of workers participate (Employee Benefit Research Institute, 2006). While retirees from active duty may also be covered under a civilian employer's pension plan, the reservist's higher tenure with a civilian employer will likely lead to greater pension benefits, because benefits usually increase with experience and those covered by pensions generally stay in their jobs longer (Gustman, Mitchell, and Steinmeier, 1994). However, frequent or long activations might disrupt and slow a reservist's civilian career progression.

Reserve duty when not activated has a relatively predictable and limited routine—a weekend of drilling each month and two weeks of training in the summer. But active-duty members frequently work long, irregular hours to maintain and repair their equipment and to prepare for and engage in inspections, exercises, training, and deployment. Many AC members spend days or weeks away from home for training, professional development courses, and exercises.

AC members and their families are relocated every few years under permanent change of station (PCS) moves, whereas reservists are not subject to such moves. A PCS move means severing ties to friends and community, and, for many members, finding new housing and changing schools for their children. Sometimes families are stationed abroad, and at other times a service member is assigned abroad on an unaccompanied tour and is separated from his family. The pattern of frequent movement also takes a toll on the earnings potential of the military spouse. Active-duty wives are less likely to be employed, work fewer weeks per year when employed (in part because of family moves), and earn a lower hourly wage, all leading to lower annual earnings on average (Hosek and Totten, 2002). By comparison, although reserve spouses must plan their family schedules and labor force participation around the reserve schedule of monthly drills and annual training, they are otherwise little affected by the reservist's regimen during times of non-activation. Their choice of geographic location, local labor market, and specific employer is little constrained by their spouse's participation in the reserves.

Reservists have been activated for military operations more frequently since the end of the Cold War than during it, and activations have increased since September 11, 2001. About 835,000 reservists have been deployed in support of contingency operations since September 11, 2001, according to congressional testimony by acting Under Secretary of Defense for Personnel and Readiness Jo Ann Rooney (2012). Many of the recent reserve deployments have been long, arguably much longer than reservists might have expected. Whether the demands of service have become more evenly distributed as a result of the increased deployment of reservists is a question that can be answered only after considering a number of factors. Actives and reservists alike signed up to serve, and today's more-frequent and lengthy deployments may be simultaneously more fulfilling and more burdensome. Lengthy absences, accompanied by uncertain and possibly abrupt changes in departure and return dates, can be expected to create more family stress than the previous pattern. The increased deployment could affect their civilian careers and earnings, while for actives there may be less disruption.[3] The reservist's absence may also adversely affect his or her employer, which might affect the employer's behavior in hiring, training, advancing, and placing reservists in positions of responsibility. The employer's eventual response to reservists' more-frequent deployment might result in a further negative effect on the reservist. The Uniformed Services Employment and Reemployment Rights Act is intended to protect the reservist's employment security, pay, and advancement opportunity from being adversely affected by activation and deployment, but the effect on reservists is ultimately an open empirical question.

More generally, the theory of insurance provides a way of thinking about how to compensate the reservist for deployment risk, including the possible adverse outcomes. The theory is to make state-contingent payments for state-contingent risks or costs. Related to this position is the intrinsic riskiness of a military career, and here economic theory suggests a compensating differential. One can think of this as an ex ante pay-

[3] Frequent deployment can disrupt training schedules, attending schools required for promotion, and access to firing ranges needed to stay qualified. However, evidence from deployments in the 1990s indicates that they did not delay promotion (Hosek and Totten, 2002).

ment for participating in a position that has higher latent chance of injury or death (as well as arduous, demanding conditions). Exactly what forms the future reserve contract will take remains to be seen, but these theoretical notions suggest that as greater burdens and risks are placed on reservists, their compensation should increase. Part of the increase should come in the form of state-contingent pay such as deployment-related pay and the provision of rehabilitative health care, disability compensation, and survivor benefits, and part should come from compensating differentials that increase reservist pay, perhaps in general or perhaps selectively depending on the incidence of the ex ante risk.

Analysis of the earnings of activated reservists in 2002 and 2003 relative to 2000 indicates that average earnings of reservists *increased* relative to what earnings would have been if reservists had not been activated (Loughran, Klerman, and Martin, 2006). Reservists who served 30 or fewer active days in 2000 and more than 30 days in 2002 and 2003 experienced a net gain of 22 percent over their base year earnings of $42,235 in 2000. Although average earnings increased with active-duty days, some reservists did experience an earnings loss when activated. Loughran, Klerman, and Martin estimated that 17 percent of those who were activated more than 30 days in 2002 or 2003 experienced a loss, with 6 percent experiencing a loss of more than $10,000. Interestingly, an even larger fraction of reservists who were *not* activated experienced an earnings loss; 40 percent of reservists who were not activated in 2002 or 2003 experienced an earnings loss. Thus, activation reduced the likelihood of experiencing an earnings loss, on average, in those years. Put differently, activation for reservists increased earnings on average, though some reservists experienced an earnings loss. Further, available evidence suggests deployments since September 11, 2001, have had little overall effect on civilian employers in terms of employment at the local level (Loughran, Klerman, and Savych, 2006).

A different perspective about equity is gained by recognizing that individuals freely choose between civilian, active-duty, and reserve careers knowing that the three choices entailed differences in commitment, risk, hours of work, and compensation, including retirement pay. This brings us to another facet of equity, which is competitiveness. In a competitive labor market, individuals are wage-takers and choose jobs that maximize utility. Likewise, a competitive market demands that employers offer wages (more generally, employment contracts) that minimize costs. The result is that individuals are paid their incremental value to the employer. If they do not like the pay or other characteristics of their job, they are free to leave and seek a job with more suitable characteristics. This is the labor market environment in which the military, as an all-volunteer force, must operate. It must offer individuals a bundle of job characteristics that attract and motivate the individuals in keeping with firm operating objectives. Competitiveness requires that the military offer no more than this. In a competitive market, then, reservists receive less retirement pay than active-duty members because reservists are willing to accept less retirement pay, either because the nature of the reserve job differs or because outside opportunities differ.

The RC has transitioned from a strategic reserve to a mix of a strategic and an operational reserve, and reservists can expect to be activated more regularly than during the Cold War. This may require an increase in reserve compensation such as higher basic pay or higher bonuses and special payments. It need not come in the form of higher retirement benefits. An equitable or fair retirement system in this view is a retirement system that, along with other forms of compensation, results in a force with desirable characteristics, without either the actives or the reservists being paid less or more than needed to achieve this force.

Recognition of More-Frequent and Longer Deployments

The increased use of reserve forces in peacetime operations, small-scale contingencies, and the operations in Iraq and Afghanistan may be changing the way reservists think about the adequacy of reserve compensation relative to the obligations of serving as a reservist. Reservists know the reserves are part of the total force and so are at risk of activation and deployment.

The congressional proposals we analyzed in Chapter Two may not be an effective response to retention and recruiting problems that might be caused by frequent and long deployments. This is because they are not targeted at personnel who actually deploy. In contrast, deployment-contingent pay is well targeted for offsetting these costs. Leading examples of deployment-related pay are family separation pay, imminent danger pay, certain places pay, and combat zone tax exclusion. Further, although the proposals might help sustain the general supply of reserve manpower, it is unlikely that they are cost-effective for this purpose relative to an increase in current pay. Current pay is received immediately by everyone and is not discounted; retirement benefits are receivable in the future (age 60) and only by those who qualify. Although we do not have a theory of how much of a pay increase should be across-the-board and how much deployment-contingent, we suggest that a formal theory would be useful in elucidating the principles.

The impact of future retirement benefits on current retention is mediated by personal discounting, and such benefits are relatively costly for the government to supply compared with current pay. Estimates of military members' personal discount rate are 6 to 10 percent per year, compared with a government interest rate of 3 to 4 percent (rates adjusted for inflation). At a 3.5 percent rate of interest, the government would need to set aside $0.71 now in order to pay $1 in ten years. But from the viewpoint of a reservist with a 10 percent rate of time preference, the present value of a dollar to be received ten years from now is $0.39. Therefore, the present cost to the government is much higher than the present value of the benefit to the reservist. A more cost-effective approach to addressing recruiting and retention problems among reservists would be through current pay and especially targeted current pay.

Ensuring an Adequate Supply of Reservists

Past studies of recruiting and retention find that military compensation—including basic pay, bonuses, allowances, and retirement benefits—powerfully influences whether the armed forces can recruit and retain sufficient numbers of qualified personnel of the right skill and experience mix. Retirement benefits play two key roles within the context of the goals of military compensation: first, to help members provide for old age; and, second, to affect the shape of the personnel force structure with respect to the distribution of personnel by rank and year of service. The military achieves the latter by providing retention and separation incentives to personnel at different ages, ranks, and years of service. This is what legislators had in mind when they established the reserve retirement system in 1948 as part of the Army and Air Force Vitalization and Retirement Equalization Act. The Committee on Armed Services in the House of Representatives detailed the purpose of reserve retirement as follows:

> The underlying purpose in writing this policy as to reserve components into law is that the retirement benefit will furnish an incentive that will hold men in the reserve components for a longer period of time. It was stressed by practically every witness who testified on this feature of the bill that the most desirable type of Reserve was a reserve of men with accumulated training. It was also pointed out that the direct monetary emoluments payable to the Reserve officers and men were so small that in many instances as the men grew older, became married, and took on family obligations, unless an additional incentive were offered them, they would drop their reserve training.[4]

Thus, another critical aspect of assessing reserve retirement reforms is their effect on recruitment and retention. We argued that discounting blunted the impact of retirement benefits on recruiting and retention relative to current compensation, but that does not mean that retirement benefits have no effect or are not a welcome part of the compensation package. The dynamic programming model provides a framework for analyzing individual choices to join, stay, and eventually retire from the reserves and how compensation policies affect these choices. Here, we discuss broad retention and force-management effects of the reserve retirement systems.

---

[4] U.S. House of Representatives (80th Congress, 1947), House Report 816 (quoted in DoD, 1988).

Retention and Sorting

Perhaps the strongest retention incentive created by reserve retirement is the incentive to earn 20 creditable YOS and thereby qualify for retirement benefits. Other things being equal, this incentive is weaker in the RC than in the AC because RC benefits are not receivable until age 60 and annual pay for non-activated RC members is lower than AC members in the same grade and year of service.

The effect of cliff vesting at 20 YOS can be seen in reserve continuation profiles. In Figure 4.1, year-to-year continuation rates at each year of service in the reserves gradually rise from around 70 to 80 percent among enlisted reservists, around 90 percent among officers with 5 YOS, and around 90 to 95 percent among reservists with 15–19 YOS. The continuation rate falls at 20 YOS; 85 percent of enlisted reservists and 88 percent of reserve officers with 20 YOS continue in that year. The continuation rate is relatively stable between 20 and 30 YOS and then falls again between 30 and 40 YOS.[5]

This brings us to another incentive effect of reserve retirement, the incentive it creates for individuals to self-sort according to ability (Asch and Warner, 1994a). The prospect of becoming vested at 20 years of creditable service will, all else being equal, have a greater retention effect on individuals who place a high probability of remaining

Figure 4.1
Reserve Continuation Rates by Years of Service, 2008

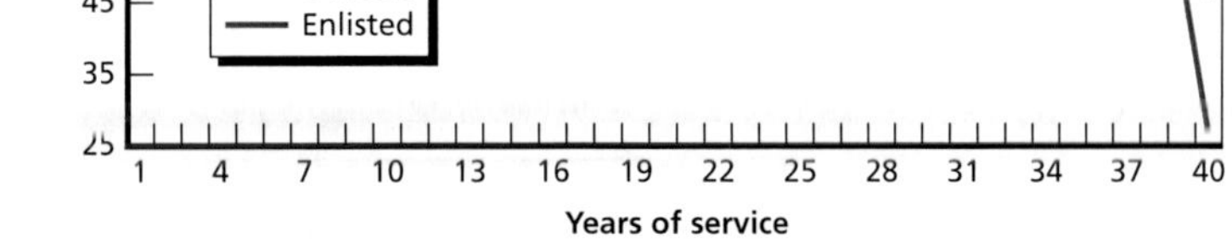

SOURCE: Official Guard and Reserve Manpower Strengths and Statistics (Office of the Assistant Secretary of Defense, Reserve Affairs).
RAND *MG378-3.7*

---

[5] Note that the continuation rates in Figure 4.1 reflect behavior for both prior and non-prior service personnel. Estimates for the dynamic retention model and the fit of the model shown in Chapter Two are for data that only include reservists with prior AC service. Thus the patterns of continuation differ between Figures 4.1 and 2.1.

in the reserves until 20 years of creditable service than on those who do not. The probability of remaining in the reserves, in turn, is in part a function of promotion prospects. On average, high-ability individuals will tend to advance faster than low-ability individuals and so are less likely to be subject to mandatory separation due to years-in-grade ("up-or-out") restrictions. In addition, their faster rate of promotion means the value of their retirement benefits will be greater, since retirement pay is a function of pay grade. Thus, vesting at 20 years of creditable service will provide a greater carrot for high-ability individuals because they are less likely to be forced out and because the value of retirement benefits increases with pay.[6]

This is precisely the type of self-sorting the military wishes to encourage. The military fills its upper echelons with individuals who move up through the ranks; there is no lateral entry into high ranks. Consequently, military compensation should encourage high-ability individuals to stay and seek promotion. At the same time, the military wants to encourage low-ability individuals to separate relatively early in their careers without necessarily forcing them to do so. Involuntary separation, while legal, has potentially high costs by adversely affecting morale—individuals may perceive the prospect of involuntary separation as risky and unfair—and encouraging individuals to lobby against the policy (Milgrom, 1988).

Although the compensation structure offers greater incentives to high-ability individuals to stay and seek higher rank, they might also have better civilian prospects for earnings and advancement. Therefore, although embedding a good incentive structure in military compensation is crucial to quality retention and sorting, the effectiveness of the incentive structure will depend on how well it measures up to outside alternatives. Large organizations, for example, also have incentive structures to keep and sort high-ability employees. The overall effectiveness of the military incentive structure depends not only on the sorting incentives but also on the retention incentives the military must pay to induce high-ability individuals to stay.

The amount the military must pay for retention depends on the correlation between ability and taste for military service. The military will be able to set a lower pay scale if ability and taste are positively correlated than it will if they are negatively correlated. But even if the correlation were zero, the military would still have to set compensation high enough to keep high-ability personnel. This could result in "overpaying" low-ability personnel if all pay were current and none deferred. An advantage of the military promotion system, which favors the retention and sorting of high-ability personnel, is that pay at lower grades can be the same for both high- and low-

---

6 A full treatment of promotion in the selected reserves extends beyond this brief discussion. It is commonly thought that promotion in the reserves often requires finding a job at a higher rank in local units because there may be no immediate opening in one's own unit. Changing units may disrupt friendships and increase or decrease travel expenses. The gains from promotion in the form of greater responsibility, greater authority, higher pay, and higher expected retirement benefits are all incentives that encourage a reservist to seek promotion despite the "transactions costs" of doing so.

ability personnel, but the value of the military career will be higher for high-ability personnel. This allows the military to avoid overpaying junior low-ability personnel in its desire to keep junior high-ability personnel.

Beyond encouraging self-sorting by ability, the structure of retirement benefits encourages all individuals to exert effort by seeking promotion (Asch and Warner, 1994a). Because the value of retirement pay increases with rank and years of satisfactory service, promotion is rewarded not only by an increase in current compensation but also by an increase in future compensation in terms of higher retirement pay and the chance to reach still higher grades, which would raise retirement pay still further.

The operation of personal policies in the RC and promotion in particular has not been a field of study in defense manpower research. Now that the RC has become a more prominent part of operation plans, this may change. Empirical analysis is needed to determine whether promotion timing, opportunity, and predictability are different in the RC than the AC. If senior outflow in the RC is slower than in the AC, the creation of vacancies in high-rank positions will also be slower. This might mean longer average time to promotion. Further, RC units are local, and vacancies may be less predictable (because of "small sample") or more predictable (because of common knowledge in the unit of the career plans of higher-ranking personnel). Some reservists have changed units to achieve promotion, going to a unit where a vacancy appears. Taken together, these factors might imply weaker incentives for self-sorting on ability in the reserves and a greater role for taste in explaining persistent participation in the RC. But again, this remains an empirical question to be studied.

Reserve Retirement and Force-Shaping in Today's Reserves

Because the reserve retirement system was designed in a different era, it is natural to ask whether the incentives it creates are desirable given the environment in which the reserves operate today. The 6th QRMC focused on reserve compensation and on retirement specifically. It indicated in its final report that maintaining an incentive to serve through 20 YOS should be a basic feature of any reserve retirement system. Deferring a portion of compensation by means of retirement vesting at 20 YOS creates an incentive for trained individuals to remain in the reserves at least to that point, thereby increasing the return on the investment made by the RC (and to some extent the AC) to train and develop its force.

The 6th QRMC and others did express concern, however, about the relatively weak incentives of the current reserve retirement system to separate voluntarily after 20 YOS.[7] On the AC side, it is clear that legislators desired a retirement system that helped maintain a young and vigorous active-duty force. Making retirement benefits

[7] There is little in the legislative history to suggest why age 60 was chosen as the age of pension receipt for reservists. The fifth edition of the Military Compensation Background Papers (DoD, 1996) speculates that age 60 was chosen because this was the minimum age at which federal civil service employees could voluntarily retire at that time.

payable immediately upon retirement is an inducement for senior personnel to leave as soon as they reach the 20-year point. For reservists, on the other hand, continued service after 20 years can only add to the value of their retirement benefit at age 60 since they do not forgo retirement pay by serving an additional year, and they do not need to start a second career. Relative to the AC system, then, RC retirement provides little incentive to separate after 20 YOS. Still, if promotions after 20 YOS in the reserves are infrequent, a reservist might choose to become a retired reservist and as such, accumulate years of service while not having to participate in monthly drills and annual active training.

At the time of the 6th QRMC, the RC complained that their force structures were more heavily concentrated in later years of service than desired. They wanted more members with 6 to 20 YOS and fewer with 25 to 30 YOS (DoD, 1988) and they attributed this imbalance in part to the absence of an immediate annuity in the reserve retirement system. The apparent reasoning was that, given a fixed hierarchy, higher senior separation increases the number of senior positions available and so increases the probability that any midcareer member will be promoted.

The results presented in Chapter Three show that the immediate annuity proposal would have achieved the force-shaping objectives indicated by the RC at the time of the 6th QRMC, while the age-55 and sliding-scale proposals would have been a small step in that direction. However, it is not clear whether the imbalance in terms of years of service poses a problem. Studies have found that the productivity of military personnel increases with their experience, although these studies focus on junior personnel rather than senior personnel and thus might not be a good guide to the gains from greater seniority in the reserves (Kavanagh, 2005). Similarly, we do not know empirically how many midcareer reservists were deterred from staying because of a lower probability of promotion; indeed, we do not know whether the increase in senior time in grade resulted from a reduced outflow, a reduced inflow (fewer promotions from the midcareer ranks), or both. Without empirical knowledge of the gains from greater seniority and the reasons for the increase in seniority, it is difficult to judge the potential benefits and cost-effectiveness of changing the current RC experience mix to a more junior—or possibly even a more senior—mix.

In the same vein, we do not know how the RC set their retention goals—or, stated somewhat differently, how authorized strengths are determined. The goals may or may not reflect a component's ongoing critical assessment of its manpower requirements given changing missions and technology. Also, it may be that the reserves could achieve their retention goals at lower cost with changes in their compensation system, including the retirement system. Research on the active-duty force indicates the feasibility of maintaining the same retention profile, increasing incentives for effort related to promotion, and yet lowering the total cost of compensation by changes that decrease deferred compensation and increase current compensation.[8] The reserves are different than the actives, and the RC retirement benefit system defers much less compensation than does the AC system (this is implied by Figure 4.1). A separate analysis would be required to determine whether similar changes in the structure of reserve compensation would be cost-effective.

As with the actives, the structure of compensation exerts a major influence on the reserve personnel force structure. Changing the structure of compensation can be expected, over time, to change the personnel force structure. Consequently, it is not only useful to ask how a given change in reserve retirement benefits would affect retention and personnel force structure and how much it would cost, but also whether the personnel force structure is being improved. With available data and models, it is easier to determine how personnel force structure will change, although this is not simple, than to determine whether the change is an improvement. Put differently, changing the reserve retirement benefit structure should not be done merely because it can be done or because it responds to concerns about perceived inequity, but also because such change is in the interest of national security.

Flexible Personnel Management

The reserve components employ individuals with a wide range of skills. While broad incentives to encourage long reserve careers may be desirable in many cases, these incentives should not interfere with the components' ability to pursue personnel management objectives. Cliff vesting at 20 YOS, for example, creates an incentive to serve at least 20 YOS but makes it difficult for personnel managers to separate personnel nearing 20 years. Such separations would seem unfair and might be perceived as prejudicial. As a result, the components may keep more midcareer members than they would in the absence of cliff vesting.

The components may also find it difficult to involuntarily separate individuals with more than 20 YOS. Here, the challenge for the RC seems greater than for the AC because, with immediate retirement benefits, the incentive to leave the AC is greater. In instances where a reserve component wants to separate reservists with more than 20 YOS, a financial incentive such as separation pay might be useful. This is also true for separating midcareer members nearing 20 years.

Given their diverse roles and missions in the total force, reserve components may want different retention profiles. Within a component, it also may be desirable to have different retention profiles for different occupations; e.g., long careers for legal professionals, pilots, and procurement specialists; and short careers in specialties that

8 Estimates of cost savings depend crucially on the assumed value of the government discount rate (see Asch, Johnson, and Warner, 1998).

demand youth and vigor. Assuming the RC will continue to have a common retirement benefit system even if it is changed from its present form, the shaping of retention profiles is probably done more cost-effectively through current rather than deferred compensation.

Future changes in force structure could affect the usefulness of retirement benefits as an incentive. There have been calls to make greater use of civilian skills and expertise; for example, the 2002 Review of the Reserve Component Contributions to National Defense argued that the transformation of the reserves should expand the capability and flexibility of the total force by taking better advantage of the civilian and military expertise of current reserve members. The review proposed that the components could access these skills by adopting a continuum-of-service approach that would allow reservists to serve in a variety of capacities from 0 to 365 days per year as needed. Here, the intent was to build in flexibility through variable-length activations and assignments. If the RC pursues this or similar approaches, it is likely that current pay, rather than retirement benefits, will be more-effective tools for the same reasons given above.

Cost-Effectiveness

Reforming the reserve retirement system need not lead to higher costs in the long run. Because personal interest rates are generally substantially higher than the government interest rate, it possible to reduce the annuity age and keep retirement costs actuarially fair from the government's perspective, but increase the present value of the retirement benefit from the reservists' viewpoint. As discussed, it is more cost-effective from an efficiency standpoint to front-load compensation in the form of pay, and to back-load compensation in the form of retirement benefits only if doing so has beneficial force-shaping and sorting implications.

Ultimately, the cost of an alternative retirement benefit system must be judged relative to its benefits to the reserve components and to reservists. If the RC wants to achieve a different retention profile, altering the retirement benefit scheme might help. Similarly, RC retirement benefits might be changed to be more competitive with retirement benefit plans offered in the civilian world, or if the RC wants greater recognition (more generous benefits) for reservists with higher cumulative deployment or longer careers. Of course, changes such as these should be assessed in comparison to alternative approaches.

CHAPTER FIVE

Reserve Retirement in the Context of Active-Duty Retirement Reform

Concern about the military's retirement system is not new. Numerous study groups and commissions have discussed reforms to the system to address problems of cost, inefficiency, lack of flexibility, and inequity since the modern retirement system was created after World War II. With the exception of the 6th QRMC, some analysis for the 9th QRMC, and more recently for the 11th QRMC, all of these past groups focused on the active retirement system, yet many of the issues raised by these groups are relevant to the reserves.

This section discusses some of the reforms recommended by past study groups.[1] We argue that achieving a compensation system that supports the seamless integration of the active and reserve components will require reserve retirement reform to be integrated with active reform, although the resulting systems will not necessarily be identical.

An Overview of Past Proposals to Reform Active Retirement

Five major issues have driven attempts at active-duty retirement reform: cost, equity, civilian comparability, force-management flexibility, and selective retention. These issues all have counterparts in reserve retirement reform.

Cost

From the Hook Commission (1948) through the recommendations of the Defense Business Board ([DBB], 2011), virtually all study groups have been concerned about the cost of providing benefits during the second-career phase of retirement (the period before the they retire from the civilian labor force), and consequently the cost of back-loading military compensation in the later part of a member's work life. Given that the congressional proposals we considered in Chapter Three would give benefits during

[1] A more detailed summary is provided in Christian (2006).

some or all of reservists' second-career phase, it is ironic that all of the past studies through the 1960s, 1970s, and 1980s recommended doing the exact opposite for the active retirement system. That is, they recommended reducing, or even eliminating, the annuity for active members during this stage of their career. Some studies, such as the Hook Commission (1948) and the Joint Pay Board in 1947, recommended eliminating the annuity during the second-career phase, and therefore making the active system look more like the reserve retirement system. Others recommended reducing the annuity; indeed, one of the major changes enacted by the passage of the 1986 revision to military retirement system, known as REDUX, was to cut benefits during the second-career phase by allowing benefits to adjust by less than the Consumer Price Index until age 62. In 2011, the most recent study group to examine military retirement, the DBB, recommended replacing the current system with a defined contribution plan that would vest early, after 3 to 5 YOS, and only begin payouts in old age, between ages 60 and 65. On the other hand, the Defense Science Board (DSB) in 2000 rejected the notion that it was excessively costly to provide benefits during the second-career phase. They argued that if the services desired a youthful organization it was appropriate to offer a benefit that helped, and induced, members to transition to civilian life when they were in their 40s and 50s. The DSB was more concerned about other issues, such as force-management inflexibility.

Equity

Nearly all of the commissions and study groups considered the issue of equity, as have many congressional proposals introduced over the years. Of particular concern was what the Joint Pay Board called the "tontine" nature of the 20-year vesting requirement: something that benefits the surviving few at the expense of the many.[2] Only a fraction of personnel stay long enough in the active component or in the active and reserve components combined to qualify for benefits. A recommendation that was often put forward was to lower the vesting requirement to 10 YOS but increase the entitlement age. The Joint Pay Board recommended an entitlement age of 62, while the Retirement Modernization Act of 1974 set it at 60. The Defense Manpower Commission (1976) recommended that the retirement annuity be paid at age 65, with a reduced annuity at age 60. More recently, the DSB in 2000 and the DBB in 2011 recommended early vesting in a defined contribution plan that would begin payout of benefits after age 60. Again, it is noteworthy that the congressional proposals seek to lower the reserve entitlement age while past proposals sought to increase the active entitlement age and to lower the vesting requirement.

---

[2] The Joint Pay Board study is cited in Asch and Warner (1994a).

Civilian Comparability

Although both the Joint Pay Board in 1947 and the President's Commission on Military Compensation in 1978 recommended providing military retirement benefits via a trust fund to which military members contribute, such contributory plans have received the greatest attention only in recent years, especially by the DSB and DBB. Driving that attention has been the dramatic growth of defined contribution (DC) plans in the civilian sector.[3] DC plans allow workers, once vested, to own their retirement assets. The portability of these assets supports the mobile workforce that characterizes the United States. It allows workers to take their benefits with them and allows employers to shed workers without being subject to the charge of opportunistic dismissal to avoid funding the retirement liabilities of the dismissed workers. It also protects workers from firm bankruptcy or under-contribution, problems plaguing DB plans. The Revenue Act of 1978 first allowed U.S. employers to offer 401(k) plans to their employees, and the number of U.S. employees participating in 401(k) plans rose from 4.4 million in 1983 to 23.1 million in 1993. By 1998, roughly half of all households were eligible to participate in 401(k) plans.[4] Poterba, Venti, and Wise (2000) used information on 401(k) participation and contribution patterns and found that 401(k) plans are likely to play a central role in providing for the retirement income of future retirees.

In FY 2000, Congress permitted military members, including reservists, to contribute to the Thrift Savings Plan, a DC plan currently offered to federal civil service employees. However, unlike their civil service counterparts, military members receive no employer or government matching contributions. Several recent studies, including those of the DBB and the DSB, have recommended that the military retirement system should include a DC plan that is vested early and funded by DoD contributions. The DBB concludes that such a system could produce sizable cost-savings. However, the DBB provided no estimates on how such a system would affect force management, specifically retention. We note that a DC plan could be provided in conjunction with a DB plan; it is not an either-or choice.

Force-Management Flexibility

Perhaps more than any other objective, force-management flexibility has been the driving issue behind calls by different study groups for retirement reform. The principal goal of military compensation is to ensure force readiness by providing a supply of members with requisite skills and experience when and where they are needed. With

---

3 In contrast, a defined benefit (DB) plan defines the retirement annuity benefit based on a formula. For example, the current reserve retirement benefit is a DB plan. The annuity benefit is described in Appendix A. In a DC plan, employees and/or employers contribute to a fund. The contributions are usually defined by a formula, but the value of the annuity depends on the level of contributions over time, and how the fund performs over time as an investment vehicle.

4 Information and discussion of the growth of 401(k) plans are found in Poterba, Venti, and Wise (1998, 2000) and Papke (1995, 1999).

respect to the retirement system, early study groups focused on the importance of incentives for separation among relatively young personnel to keep the armed forces "alert and vigorous" according to the Hook Commission in 1948.

Selective Retention

The main concern regarding force-management flexibility, however, was the greater-than-desired conformity of career lengths among military members, regardless of their occupation or specialty. The one-size-fits-all aspect of the AC retirement system and the strong pull of the 20-year vesting requirement for those with less than 20 YOS creates similarity in the experience mix of personnel across occupational areas and hampers the ability of the services to manage areas differently. Later commissions and study groups understood the value of giving the services the flexibility to manage skills separately. The Retirement Modernization Act of 1974 proposed a system of voluntary and involuntary separation payments. The Defense Manpower Commission in 1976 proposed a point system for the receipt of annuity benefits for active members, with those in combat roles having the ability to earn points at a faster rate than those in non-combat roles. During the defense drawdown in the early 1990s, following the end of the Cold War, the voluntary separation incentive and special separation benefit were used by the services to target the separation of personnel in specific occupations in specific years of service and pay grade groups. These incentives were tremendously successful in achieving dramatic reductions in end strength, especially in the Army and Air Force, while providing members with a benefit that eased their transition to civilian life. They were also highly effective at targeting the separation of lower-quality personnel in terms of Armed Forces Qualification Test scores and high school diploma status (Asch and Warner, 2001).

To address the lack of flexibility embedded in the retirement system, some recent study groups have recommended a permanent system of separation benefits, including the 2000 DSB, the Defense Advisory Committee on Military Compensation (2006), and the 10th QRMC (2008). For example, the DSB recommended that the benefits would be an annuity based on the current retirement system formula (High-3) and would be received from separation to age 62. At 62, the individual could draw from his or her thrift savings plan fund. The separation benefits could be used to achieve differing career lengths in different skill areas. Areas where a shorter career is sufficient, such as combat arms, would receive the benefit early on when they are younger, while areas where longer careers are desirable, such as computer programmers, could begin receipt at older ages. In this way the retirement system would achieve its twofold purpose: helping members to accumulate savings for retirement (via the thrift saving plan vested early and paying benefits at age 62) and providing flexibility for the services to manage their personnel (via a system of separation annuities paid from the date of separation until age 62).

Relevance of Proposals to Reserve Retirement Reform

The issues surrounding the active retirement system are relevant to those related to reserve retirement reform, but there are differences. While equity is of concern for the reserve system, the issue is equity vis-à-vis the active system and the integration of the two systems, not necessarily just equity vis-à-vis those who do and do not reach the 20-year vesting point. Still, the 20-year vesting rule is a component of the reserve system and therefore, it is arguably an equity issue in the RC as well as in the AC. Similarly, management flexibility is of concern in the reserves, but revolves around assuring the retention of trained personnel while preventing "superannuation" caused by weak incentives to leave the reserves after reaching 20 YOS. In the active components, the issue revolves around the uniformity of career lengths and the one-size-fits-all career produced by the active retirement system. Nonetheless, uniformity of career lengths across skill areas within components is relevant to the reserves as well. Comparability with the private sector and the call for 401(k) plans with matching contributions is another issue that has sparked debate about the active system, yet has been muted in debates about the reserve system.[5] Again, this issue is relevant because, like the active system, the reserve system is a defined benefit, not a defined contribution plan, and member contributions to their TSP funds are not currently matched by DoD.

That the issues surrounding retirement reform for the active and reserve systems are relevant but not always the same suggests that reforms for each must work in concert to achieve their respective personnel goals. The idea of total force management, along with the seamless integration of the active and reserve components, has received considerable attention in the past decade, and especially since September 11, 2001. Although the retirement systems for the components need not be identical, alternatives to reform either system should be judged in terms of how they support the total force. Therefore, proposals that call for a 401(k)-type plan, e.g., a TSP and system of separation pay, should be assessed not only in terms of their effect on active personnel outcomes, such as retention and cost, but also on reserve outcomes, such as reserve affiliation, retention, and cost. Similarly, proposals such as the congressional bills analyzed in Chapter Three that seek to reduce the age of entitlement for reserve retirement benefits should factor in the effects on active members, including in terms of equity as discussed in Chapter Four.

[5] An exception is the debate in 1999 about whether reservists should be allowed to participate in the Thrift Savings Plan (TSP) that was to be provided in FY 2000 for active members. The concern of the TSP investment board was that participation and levels of contributions of reservists would be low, yet the cost of administering their fund accumulations would be high. Ultimately, it was decided to include reservists in the legislation that permitted military members to contribute to the TSP (see Asch and Warner, 2000).

CHAPTER SIX

Implementation and Obstacles to Reform

Despite the many recommendations to change the active-duty retirement system that have been made over the years by studies and commissions, few of the recommendations have been adopted and in fact the system has changed very little. Changes that were made in 1981 and 1986 (see Appendix A) did not respond to the primary concerns expressed by the study groups about equity, flexibility, and the cost of back-loading the system due to the 20-year cliff-vesting provision. As discussed in a later subsection, they were mostly viewed as cost-cutting moves. The change in 2000 restored the pre-1986 retirement system but also gave service members the choice of an alternative, namely, a $30,000 bonus in the 15th year of service tied to a commitment to stay in service for another five years and then retirement benefits at the REDUX rate.

Given the lack of success past study groups have had in effecting change, despite their distinguished members and thoughtful recommendations, it seems that identifying and overcoming the obstacles to reform is as essential as developing a set of recommendations if reform is to occur in either the active or reserve retirement system, or both.

What are the obstacles? Aside from cost, the key obstacles have to do with the lack of consensus for change. DoD cannot unilaterally change the military compensation system the way a private sector organization might; Congress can only make such changes in the form of legislation that must be signed by the President. Members of Congress generally respond to concerns of constituents and lobbying groups, such as service members and retirees from the armed services, or to readiness concerns articulated by the armed services during congressional testimony. The DoD also provides testimony and advises the White House on military compensation issues.

With regard to retirement reform, fundamental change has not been a priority of the services. From the perspective of the services, both the active and reserve compensation systems (including their respective retirement systems) have, by and large, performed well since World War II in delivering stable supplies of personnel to the armed services in terms of the number, experience, and skill mix needed. Though shortfalls have occurred, and quality has varied as external economic conditions have changed and military compensation has risen or fallen relative to civilian opportunities, shortfalls have been successfully addressed with the use of special and incentive pay, (e.g., reenlistment bonuses and continuation pay), as well as changes in military pay. Furthermore, none of the services has specified an unmet requirement for flexibility in managing personnel. That is, the commissions and study groups have argued for greater flexibility, yet the services have not clamored for it. The supply of service members by year of service appears to drive the requirement for experience. Thus, the services' requirements for personnel at each year of service seem to reflect the manpower willing to serve at that year of service under the existing retirement and compensation system.[1] Furthermore, the commissions have not provided the services with quantitative evidence on how the reforms would improve flexibility or defense capability and readiness. Lacking such evidence but showing an ability to adapt to the retention profile under the current retirement system, the services have not called for any major retirement reforms in their congressional testimonies, such as those that would eliminate the 20-year cliff-vesting rule.

In addition, the call for reform has met with resistance among some in the retiree and veterans' populations as well as among current service members. Revamping the retirement system raises fears of broken trust, benefit cuts, and an open door to future rounds of disruptive and demoralizing changes. This could erode morale and perhaps reduce defense capability.

Another facet of the lack of consensus obstacle is disagreement about the main objectives of the retirement system. Dr. Kenneth Coffey of the then–General Accounting Office summarized this issue well in his congressional testimony in 1983:

> In our view, these previous attempts at retirement reform have failed, in part because all parties concerned—i.e., the military services, top civilian Defense officials, the Office of Management and Budget, and the Congress—have not agreed on what the controlling objectives of the retirement system should be and what fundamental compensation principles should guide the work of study groups. (This is not to say that the study groups were not guided by specific objectives, but rather that all parties affected had not bought into the specific objectives or principles adopted by the study group.) Some critics of the present system assert that the primary or controlling objective of the non-disability retirement system should be the care of those who, because of age, are no longer capable of performing military duties. Defenders of the current structure contend that the system is not, and should not be, primarily an old age pension program, but rather the retirement system should serve as a force-management tool.[2]

As further noted by Coffey in his testimony, retiree groups and other like-minded organizations saw the retirement benefit as a reward in recognition of military service

---

[1] The relationship between compensation, manpower supply, and the determination of requirements is elaborated elsewhere (Asch and Warner, 1994a; Asch and Hosek, 2004).

[2] See Coffey (1983), p. 64.

performed by members and paid at the end of their work-life to help qualified veterans transition out of the labor force. From this perspective, any reduction in retirement benefits was considered breaking faith with the implicit contract with members, veterans, and retirees. In contrast, managers and individuals such as those who formed and often participated in the various retirement and compensation study groups and commissions saw the retirement benefit as an incentive to shape the age and experience mix of the armed forces. From this perspective, retirement had to be compared to basic pay and other compensation elements such as special and incentive pay in terms of cost-effectiveness, force-management flexibility, and efficiency.

The next section sketches an economic model of retirement reform where efficiency in the provision of national defense and the interests of taxpayers is traded off against the interests of a specific group, namely service members and retirees, whose members are heterogeneous and where some members prefer the status quo and would like to block reform. A technical presentation of the model is given in Appendix D. The theory provides insights on the factors affecting whether reform occurs and strategies for increasing the likelihood of successful reform, such of the introduction of menus of reform options.

An Outline of a Theory of Successful Reform

Demange and Geoffard (2006) developed a theory of compensation reform under political constraints and analyzed the issue in the context of managed health care in Europe, where reformers would like to improve efficiency by introducing physician payment schemes that offer incentives to provide high-quality care for the same cost or to provide the current quality of care at a lower cost. The political constraint is that physician groups prefer to block reform efforts that improve efficiency and patient welfare, because those reforms would reduce the welfare of some physicians. They develop a set of mathematical conditions that indicate under what circumstances successful reform is likely.

In this section, we apply the Demange and Geoffard model to the question of politically feasible military retirement reform. While it is possible to write down the equations for the model for our application, additional work was necessary to embed their model into the dynamic programming model. Once that was done, we could examine the effects of alternative retirement reform proposals not only on retention and cost but also their political feasibility. However, integrating this modeling into the DRM is beyond the scope of this report. Here, we simply apply the main components of the Demange and Geoffard model to retirement reform. This application is presented formally in Appendix D. We present an overview of the application in this subsection.

In the retirement reform context, we consider a triad: the general public, who is the consumer of national security and who is also the taxpayer; service members, who "produce" national security and who must be compensated; and Congress.[3] The general public pays taxes and is subject to a spectrum of national security threats that are uncertain and assumed to be random and exogenous shocks. Service members are assumed to be heterogeneous in their taste for service and can vary their effort level and decide whether to stay or leave.[4] National defense depends positively on effort and experience levels, but is subject to the random exogenous shocks. Congress is assumed to be an intermediary that collects taxes from the general public and disperses them to service members in the form of pay. The key problem faced by Congress is that it is not feasible to compensate service members directly on the basis of defense output because of difficulty of defining measures of output. Consequently, they are paid on the basis of inputs, namely effort and/or years of military experience or career length. Service members are assumed to care about national defense because of motives such as patriotism and to be willing to provide effort even with zero financial incentives. Similarly, the general public values national security and is therefore willing to pay taxes to support the provision of it.

The structure of the model is as follows. We identify the level of service member effort and career length as well as the military compensation scheme that would be the most efficient in terms of maximizing total expected welfare, i.e., we identify the scheme that would make everyone the best off jointly. This efficient compensation scheme is called the first-best scheme. If the first-best scheme has not been achieved, but some other suboptimal status quo has been achieved instead, we then ask if reform is feasible. A reform is a new scheme that would increase efficiency over the status quo (or not reduce it), and improve the situation for a large enough proportion of service members that they would not block reform. We then consider the circumstances when service members would not block reform. We provide more detail about these steps.

Each member of the triad maximizes its expected value. For service members, this involves choosing the effort level and career length that maximizes the difference between the military compensation they receive and the opportunity cost of their effort and time, plus the value of their patriotism from serving in the military. The general public maximizes the expected value of national security minus the taxes paid. We assume Congress has a costless role and its expected value is the sum of the taxes

---

3 In reality there are more than three groups, including the individual AC and RC and OSD.

4 In the Demange and Geoffard model, producers vary in terms of talent, not taste. However, because we have already modeled heterogeneity in terms of taste for military service in the context of the dynamic retention model, it is convenient to continue to do so, although the extension to talent can be made—and has been elsewhere (Asch and Warner 1994a, Hosek and Mattock, 2003). Also in the model, producers choose their effort levels only, not their stay or leave decision. We added the retention decision as it would be part of any extension of the model to the DRM.

collected from the general public minus the sum of the expenditures dispersed to service members in the form of military compensation.

Total welfare for the triad is the sum of their expected values, and the efficient or first-best compensation scheme is the one that induces service members to choose effort levels and career lengths that maximize total welfare. In cases where the first-best compensation scheme has not been achieved, a reform would improve efficiency and reduce the taxes paid by the general public to achieve a given level of national defense. That is, a new scheme would induce the same effort and retention from service members but at less cost to the taxpayer. The question is, how can a reform be achieved when some service members are reluctant to be hurt by the reform, and therefore would politically block it?

A reform, i.e., a new compensation scheme, is politically feasible if:

- It balances the budget. The sum of taxes collected from the general public equals total expenditures paid to service members.
- A large enough proportion of service members accept the reform, i.e., the new scheme is preferred over the status quo scheme by a large enough proportion of members.

Here, we consider the member at the margin of support, defined by his or her taste for service. For example, if a 90 percent majority is needed for the reform to be acceptable, then the marginal member is the one member with taste at the 90th percentile. Ninety percent of the members would prefer reform more than the marginal individual, and 10 percent would prefer the status quo more than the marginal individual.

A reform will only be politically feasible if members are "bought off." In economics jargon, this is called a compensating variation. When members are hurt by the reform, an amount is paid that causes the financial well-being of the marginal member to be same under the status quo as under the reform. The marginal member determines the size of the buyoff or compensating variation because he or she determines the proportion of members needed to make the reform politically feasible. Thus, if 60 percent of the members must be at least as well off by the reform for it to be politically feasible, the reform must involve giving *all* members an increase in compensation that makes the marginal member equally well off financially between the status quo and the reform. The fact that some members are part of the 40 percent who would be willing to accept a smaller buyoff or perhaps even be better off under the reform means they are receiving an economic rent, or a payment that is over and beyond what is necessary to induce them to overcome their objections to the reform.

At some point, the total cost of the buyoff to members exceeds the efficiency gain to taxpayers of lower taxes or better national defense, at which point the reform is no longer sensible. It is useful to consider the conditions when reform will occur. Political reform is more likely to be successful when the total cost of the buyoff is smaller. This occurs when service members have less political power, are more homogeneous, and their effort is more responsive to financial incentives.

When members have less political power, a smaller buyoff is required to buy the marginal member to accept the reform, and thus, the total cost of the buyoff is smaller. When members have more homogeneous or similar tastes, a small increase in the buyout can have a large increase in the fraction who no longer block reform because preferences are bunched at specific points in the distribution of tastes. Heterogeneity has a positive effect on the difference between the average effort across all members and that of the marginal member, thereby increasing the likelihood that members will require a larger buyoff for feasible reform. Finally, the more sensitive that member effort (retention) is to economic incentives, the smaller the buyoff they require.

It is useful to consider the case when it is politically infeasible to offer a single new reform, but politically feasible to offer a menu of reforms that appeals to different types of members. A menu might be feasible because the schemes may be cross-subsidizing, resulting in a balanced budget across schemes, even when schemes do not balance individually. The menu has the advantage that members can voluntarily self-select and choose the scheme that is most preferred, providing the highest expected value. Thus, a menu may achieve reform when a single proposal does not. More generally, when voluntary choice is embedded in the reform effort, implementation is more feasible and more likely to occur.

Applying the Theory to Military Retirement Reform

Given the implications that reforms are less likely to be successful if those who block them are politically more powerful, more heterogeneous, and less responsive to incentives, it is useful to reconsider past attempts at military retirement reform. First, there are heterogeneous constituencies of active and reserve personnel, enlisted and officer, junior and senior personnel, in addition to military veterans and retirees and individual service differences. Different groups of personnel may create alliances favoring or opposing different reform proposals. Second, these groups' levels of influence differs but are widely regarded as politically powerful; veterans, military retirees, and reservists have political lobbying groups that represent their interests to Congress. Furthermore, public interest in the welfare of military members is high, especially in light of their sacrifices in support of operations in Iraq and Afghanistan, meaning that lobbying groups can count on the power of the media. Third, a related point is that military compensation is tremendously costly. Given the lack of service demand for flexibility and the lack of evidence of increased defense capability or efficiency as a result of reform, any buyoff of constituency groups as a tradeoff for improved capability has been viewed as prohibitively costly, i.e., costly relative to its perceived benefit. Consequently, no major reform in military retirement has occurred since World War II.

The theory also provides the key elements for developing a strategy to address the obstacles to reform:

1. Identify the key constituents involved in the reform process, especially those who would block it.
2. Assess the benefits and losses to members from reform and the timing of those benefits and losses.
3. Incorporate the buyoff necessary to ensure the feasibility of reform.
4. Consider a menu of reforms to allow choice.

Assessing the benefits and losses to members (No. 2) is particularly important because those aspects often determine whether a group will block reform. Furthermore, past study groups recommending reform have tended to ignore the issue of losses and benefits to specific groups, and have instead focused on only efficiency issues such as retention and cost.

Reconsider the examples of the 1986 REDUX retirement reform and the return to its pre-1986 predecessor in 2000 due to the TRIAD legislation. REDUX decreased retirement benefits for members who entered after it was enacted in August 1986. In the past, members who served at least 20 years received at least 50 percent of pay and had benefits that were fully protected from the erosive effects of inflation. As described earlier and in Appendix A, REDUX reduced the percentage to 40 percent for those serving 20 years, ramping it up to 75 percent for those serving 30, and did not offer full inflation protection. It was hoped, according to some in Congress, that the difference between 40 percent at 20 years and 75 percent at 30 years would provide members with a stronger incentive to stay for a full 30-year career. Others saw the change as budget-driven and a means of reducing the cost of military personnel.[5] Thus, the driving forces behind REDUX were force-management improvement and efficiency gains, and the concerns of taxpayers. But the concerns of some producers were also considered; both existing service members and retirees were bought off by the legislation in the sense that they were grandfathered under the pre-1986 system, as the theory would suggest.

Despite the concessions to taxpayers and existing members in 1986, REDUX did not "buy off" a sufficient number of supporters. REDUX was by and large repealed by the TRIAD legislation that began in 2000. Why? First, REDUX broke faith with post-1986 members by reducing their benefits. Second, it did not improve management flexibility or the cliff-vesting and immediate benefit features of the 20-year system.

---

[5] An interesting analysis of the politics of the REDUX reform legislation and the later TRIAD legislation is provided in Freedberg (1999). Rep. Ike Skelton (D-Mo.) is quoted as saying, "The purpose [of REDUX] was to cause people to stay the full 30 years..." Freedberg also quotes a "Hill staffer" as saying that REDUX "was basically a budget-driven exercise."

While it was true that REDUX provided more powerful incentives for those members who made it to 20 YOS to stay until their 30th year, it also reduced the incentive to stay until 20 YOS. Indeed, average man-years were expected to decline under REDUX (Asch and Warner, 1994b). Furthermore, no evidence was ever offered of how REDUX would improve defense capability, other than to reduce retirement costs and midcareer retention. Thus, when military recruiting and retention showed signs of duress in the late 1990s and military personnel who had entered after 1986 began to recognize that they were under a less-generous retirement plan than their predecessors, the debate centered on intercohort equity, not on the efficiency of sustaining REDUX. The service chiefs were concerned about the potential negative impact of eroded morale due to a less-generous retirement system, not on whether the previous retirement system was preventing them from meeting a requirement for management flexibility, a requirement they had never articulated. Consequently, there were no consumers rallying support behind REDUX.

Consistent with the model, the TRIAD legislation offered a menu to post-1986 members, allowing them to choose between staying under REDUX and receiving a $30,000 Career Status Bonus at the 15th year of service for committing to stay a full 20 years, or taking the pre-1986 retirement system covering their more-senior colleagues that provided 50 percent of pay at 20 YOS. The lump-sum offer was derived from the experience of the drawdown when service members opted for smaller lump sum payments over more-generous annuities. Available evidence from the Navy indicates that a significant number of members chose the lump sum. Among sailors entering between 1986 and 1989 and making their retirement system choice between 2001 and 2004, 49 percent of enlisted personnel and 13 percent of officers chose REDUX and the $30,000 bonus over the higher annuity option. The higher rate among enlisted personnel probably reflects their higher personal discount rates (Warner, 2005).

A Case Study of Successful Reform: The Federal Employees Retirement System

There is a case of a successful overhaul of a retirement system in the federal government, and it is an instructive case study of feasibly achieving retirement reform as outlined by the Demange and Geoffard framework. The case is the transition from the Civil Service Retirement System (CSRS) covering federal civil service employees beginning in 1921 to the Federal Employees Retirement System (FERS) that went into effect January 1, 1987.

The CSRS is a defined-benefit retirement plan that, because it was introduced prior to the 1936 advent of Social Security, does not include Social Security coverage. Many federal employees also had sufficient years in covered employment, however, so they also qualified for Social Security benefits, and in some cases even qualified for military retirement benefits. Concerns about "double dipping" and "triple dipping" led to threats of cuts in the CSRS benefit. There was also concern about a substantial unfunded liability that CSRS generated. CSRS is funded from contributions as a percent of pay—generally 7 percent each from the employee and the employing agency. To avert CSRS cuts, the new, three-part retirement system called FERS was created, which consists of Social Security, a relatively small defined-benefit plan called the basic plan, and the defined contribution plan, the Thrift Savings Plan (TSP). The TSP for federal employees provides automatic government contributions for each employee of a 100 percent match for a contribution of 1 to 3 percent and a 50 percent match for the next 2 percent. An individual can contribute more, up to an Internal Revenue Service maximum. The TSP has various investment vehicles, and the value of the retirement benefit depends on the level and timing of contributions, as well as the performance of the investment fund over time. As a DC plan, the TSP is portable once a member is vested. In establishing FERS, Congress placed a new risk on employees covered by FERS: Their TSP fund accumulation might fall.

The implementation of FERS was particularly successfully and ensured that existing employees and new hires were on average made either better off or no worse off by the transition. There are three groups of employees to consider:[6] new employees hired after FERS was created, existing employees covered by CSRS who were hired before FERS was created, and rehired employees who were initially covered by CSRS but were rehired after the start date of FERS.[7] New employees hired after December 31, 1983, were automatically placed in FERS. Under a range of assumptions, Asch and Warner (1999) find that net expected lifetime earnings and retirement wealth is greater for new hires covered by FERS than by CSRS. The analysis accounts for contributions to retirement, so the comparison of earnings is also relevant.[8] Employees with less than five YOS were also automatically placed under FERS and, for these employees, FERS

[6] We do not analyze the third group's retirement system, CSRS-offset. See Asch and Warner (1999).

[7] An additional complication is that FERS began on January 1, 1987, as noted, but covered employees hired three years earlier, after December 31, 1983. While the new retirement system was being developed, these employees were covered by an interim retirement system known as CSRS-interim, later called CSRS-offset, a plan consisting of CSRS plus Social Security. Those with less than five YOS were transferred to FERS beginning January 1, 1988. Those who were hired with more than five YOS were allowed to maintain CSRS-offset coverage.

[8] The greater generosity of FERS is not due to high stock market returns and the implied beneficial effect on TSP returns. The base analysis assumes a 6 percent real growth rate in TSP returns. Rather, net expected wealth is greater under FERS because of a combination of factors including accumulation of benefits from the three components, the opportunity to earn an average return from the TSP (which can protect the fund accumulation from the erosive effects of inflation over time), Social Security coverage, and the lack of a windfall elimination provision for those covered by FERS. The latter factor can have a detrimental effect on net wealth for those covered by CSRS and who enter the civil service at relatively older ages. Social Security includes a windfall elimination provision that partially deducts the employee's Social Security benefit for his or her CSRS annuity. The deduction is larger for those who enter the civil service at older ages because they usually have some Social Security covered employment.

was also found to be a more generous retirement plan on average. Thus, the first key group of constituents—new hires and those involuntarily placed in the new system—was "bought off" with a more generous system, on average, that included a portable defined-benefit fund as well as Social Security.

The second (and by far the largest) group, existing employees covered by CSRS with more than five YOS, was made no worse off by the transition because those individuals were given the choice to voluntarily switch to FERS during an open enrollment window from 1987 to 1988. Additional opportunities to enroll in FERS were given in later years as well. Thus, they could opt to stay in their current plan or take the new system. Interestingly, although existing employees were permitted to switch, few did. Asch and Warner (1999) compute the financial incentive to retire at different ages as well as the incentive to switch from CSRS to FERS and find that employees who were more senior had no financial incentive to switch. Alternatively, switching to FERS provides a larger differential if switching occurs early in the career.

The lesson here is that existing employees covered by CSRS were generally content to stay put. Furthermore, the menu that existing employees faced did not just include staying with CSRS or switching to FERS, it included a third option. CSRS members were permitted to open and maintain TSP accounts. The government would not contribute to these accounts, unlike those of their FERS-covered colleagues. However, CSRS-covered employees could put aside tax-deferred savings for retirement.

The aspects of the FERS case study that are elements of the theory of successful reform are:

- Existing members were given a menu of choices that included the current retirement system and the new system, thereby ensuring they were at least as well off under the reform.
- Existing employees were given numerous opportunities to switch to the new system.
- The new system was more generous on average than the old one for new and junior employees, but less generous for existing senior employees, on average. Alternatively, incumbent employees could be grandfathered in.
- The new system had portable components (TSP, Social Security).
- Perhaps the most instructive lesson is that it is possible to reform a federal retirement system and do it right from the employees' perspective.

CHAPTER SEVEN

Policy Implications and Conclusions

Our main conclusion is that congressional proposals to reduce the age at which eligible members may begin receiving retirement benefits are not cost-effective means of sustaining or increasing RC retention. The age-55 and sliding-scale proposals we analyzed would cause small changes in reserve retention, tending to increase it overall, but the cost per member would increase. The immediate annuity has large effects on the force profile, increasing retention prior to 20 YOS, as more RC members participate to qualify for immediate RC retirement benefits, but reducing retention after 20 YOS as more RC members leave and claim immediate benefits. The decline in senior retention more than offsets the increase in midcareer retention in many cases, and RC force size decreases. Total costs per RC member increase under all three proposals, and they increase substantially (by almost 25 percent) under the immediate annuity proposal. This proposal pays benefits earlier and therefore for more years, and total benefit costs increase because more reservists stay until retirement.

Because the proposals' effects on reserve retention vary by age group—with the age-55 and immediate annuity proposals having negative effects on senior retention and positive effects on midcareer retention—the ideal would be to judge the effects relative to the reserve components' manpower requirements. But we do not know those requirements and we recognize that they might change as the component matures as an operational reserve in addition to a strategic reserve. Components might not want senior reservists to retire earlier, or alternatively, even under today's RC compensation system, some components might have a more senior force than they desire.

We discussed the proposals in the context of active/reserve equity and improved reserve personnel management. One rationale for decreasing the reserve retirement age is that reservists receive much less in retirement benefits than do active-duty members. Reserve benefits are based on pro rata years of service and are not paid until age 60. But the equity case for the reform proposals is not clear; active and reserve service place different demands on members so it would not be equitable to treat reservists and active-duty members identically. Active-duty members are subject to PCS moves, cannot hold a civilian job and receive its pay and benefits, and face the readiness and deployment demands of full-time active service. Further, one purpose of active-duty retirement benefits is to help the retired member establish a civilian career, whereas reservists typically already have such a career as well as a retirement benefit plan with their employer. Finally, the choice of age 55 and the ages in the sliding-scale formula are ad hoc. Decreasing the age of benefit receipt is a step toward equity with the AC's immediately available benefits, but only a small one.

In terms of force management, lowering the retirement age does not improve management flexibility or provide the services with a tool to allow more-variable career lengths by occupation. The reserve retirement system encourages similarity in careers. The congressional proposals would reduce retirement ages but not increase variability in career length, and would decrease cost-effectiveness relative to the current system by increasing retirements and retirement costs along with the cost per member of sustaining the current force.

Given the reserves' expanded role in defense, we also considered reserve retirement reform with respect to active-duty retirement reform. Study groups and commissions have considered how to improve the active-duty retirement system, but only the 6th and more recently the 11th QRMC have addressed reserve retirement. Critics of the active system have focused on cost, equity, and the dampening effect on force-management flexibility. These issues are also relevant to the reserve retirement system.

Many proposals have been put forward to change the active-duty retirement system, yet only slight alterations have actually been made in the three times the system has changed in 1980, 1986, and 2000. Past recommendations for retirement reform, which focused on force management and cost-effectiveness, may not have paid sufficient attention to political barriers and specifically to the concerns of members, veterans, and retiree groups regarding equity and breaking faith with more-recent generations of qualified Americans who choose to serve in the military.

Our assessment suggests that a menu of member options can be a powerful tool to overcome obstacles to reform. Current members could be given the choice of staying in the current retirement system or joining the new one, and the choice might be offered over a period of time, say five years, rather than as a one-time deal. New entrants to the AC or RC and reentrants with few years of service might be placed under the new system. Further, the new system must offer a benefit that is attractive to gain enough members' support overall. A choice between the current system and the new system will leave current members at least as well off as they are, and the benefits under the new system must be high enough to recruit and retain new members (other things constant) and to gain the support of service organizations—and perhaps even service members, acting as stewards for the welfare of future service members.

The dynamic retention model provided an evaluation of retention and cost but was not used to assess proposals with respect to political acceptability. While we have not yet extended our dynamic programming model to compute the distribution of expected gain (or loss) under retirement reform, it is possible to do so. We think this is a fruitful area for future research. Further, it would be helpful to the policy discussion to have more information and studies on AC and RC manpower requirements, the gains from greater flexibility in designing careers and managing personnel, the value of offering a menu of choices, the value to service members of a system that would qualify more members for retirement benefits, and the form and level of current compensation given a future with higher expected reserve deployment.

APPENDIX A

The Reserve and Active-Duty Retirement Systems

Members of the reserve components who accumulate 20 years of creditable service with the last eight years of qualifying service in the Ready Reservess are entitled to receive retirement pay beginning at age 60.[1] No retirement pay is provided to members separating from the reserves with less than 20 calendar YOS. Retired pay at age 60 is calculated based upon years of creditable service (YCS) when transferred from the Ready Reservess and basic pay as calculated under one of several methods discussed below:

$$Y = YCS \times 0.025 \times BP \quad (A.1)$$

where Y is monthly retired pay and BP is monthly basic pay. Roughly speaking, years of creditable service are a pro-rated number of calendar years of service. Specifically, years of creditable service are calculated by dividing a reservist's accumulated retirement points by 360. Retirement points are earned as follows:

- One point for each day of active-duty service
- One point for each period of inactive-duty training (IDT)
- One point for each day in funeral honors duty status
- One point for each accredited three-credit-hour correspondence course satisfactorily completed
- Fifteen points for each year of active status membership in a reserve component.

Under law, reservists may accumulate no more than 90 inactive-duty points (annual membership, IDT, and course credit points) and a total of 365 active and inactive-duty points combined in a single year. The limit on inactive-duty points has been relaxed in recent years. Prior to retirement years ending September 23, 1996, annual inactive-duty points were capped at 60. This increased to 75 points for retire-

[1] The Ready Reserves encompasses the Selected Reserve, the Individual Ready Reserve, and Inactive National Guard. It excludes the Retired Reserves. Between October 1994 and September 2001, the number of qualifying years was reduced from eight to six.

ment years ending between September 23, 1996, and October 30, 2000, and stands at 90 points for years after October 30, 2000. There is also a career limit on retirement points of 10,950 or 30 YCS. A minimum of 50 points must be earned in a year for that year to qualify as a creditable year and count toward meeting the 20 calendar YOS minimum for vesting in retired pay. The average enlisted reservist separating from the Ready Reserves in FY 2000 had accumulated 2,984 retirement points over 25 calendar years of active-duty and reserve service (Asch, Hosek, and Loughran, 2006). The average reserve officer retiring in FY 2000 had accumulated 3,585 retirement points over 27 calendar YOS. Median retirement point accumulation per year among all reservists totaled 77 for enlisted members and 79 for officers in FY 2000.

The computation of *BP* depends on when the reserve member first entered military service and whether he or she transferred to the Retired Reserves upon separating from the Ready Reserves. For members entering prior to September 8, 1980, *BP* is the basic pay in effect for a given rank and calendar years of service when the member first begins to receive retired pay. A member can continue to accumulate calendar years of service (i.e., longevity) for the purposes of computing *BP* if he or she transfers to the Retired Reserves after separating from the Ready Reserves. Consequently, individuals who separate from the Ready Reserves prior to reaching the highest level of basic pay at that rank can increase *BP* by remaining in the Retired Reserves. Members of the Retired Reserves are not required to participate in drilling or training but can be called to active duty without consent in the interest of national defense. They receive no compensation and do not accumulate retirement points.

For members who enter on or after September 8, 1980, *BP* is computed as the average of the highest 36 months of basic pay ("High-3 averaging"). For reservists who transfer to the Retired Reserves, High-3 averaging takes place over basic pay in their last 3 YOS in the Retired Reserves (typically, ages 57–59). For reservists who end their affiliation with the Reserves upon separation from the Ready Reserves, *BP* is calculated over their last 3 YOS in the Ready Reserves. This distinction creates very strong incentives for reservists to remain in the Retired Reserves until age 60 so that *BP* at age 60 reflects basic pay adjustments (year-to-year increases) subsequent to separation from the Ready Reserves as well as any increases in pay from longevity. There is no incentive to delay retirement beyond age 60. All members below Major General must separate by age 60 and limits on calendar years of service may force some reservists to separate before age 60. Retired pay beginning at age 60 for all members is adjusted for inflation using the Consumer Price Index (CPI) for urban wage earners.

The most significant difference between the active and reserve retirement systems is that active-duty members with 20 or more calendar YOS begin receiving retirement pay immediately upon separating from the active-duty force instead of at age 60 as under the reserve retirement system. There are also differences in the formula used to convert YCS and basic pay to retirement pay, the most important being that YCS equal years of calendar service for active-duty members. The 2008 Defense Authorization

Act allows reservists to begin drawing retirement pay three months earlier than age 60 for every 90 days of active duty under certain mobilization authorities in support of a contingency operation, down to age 50. The law only applies for deployment time service after January 2008.

There are three different systems under which active-duty retirement pay can be calculated. For members entering military service prior to September 8, 1980, active-duty retirement pay is computed using the formula in Equation A.1 and *BP* is basic pay on the date of separation.[2] For members entering military service between September 8, 1980, and July 31, 1986, *BP* is calculated under the High-3 averaging method. Under both of these systems, annual retirement pay is adjusted using the CPI urban wage earners series.

Active-duty members entering service after July 31, 1986, choose between two retirement systems in their 15th year of service. The first system has a multiplier of 0.025 per year, so retirement benefits are 50 percent of High-3 basic pay at the 20th year of service and increase to 75 percent at the 30th. The second system is REDUX. Under it, active-duty members receive a $30,000 Career Status Bonus in the 15th year and their retirement pay is calculated by the following formula:

$$Y = [0.40 + 0.035(YCS - 20)] \times BP \qquad \text{(A.2)}$$

where *BP* is again High-3 basic pay. Between the year of retirement and age 62, retirement pay under REDUX is adjusted by CPI minus one percentage point. At age 62, REDUX makes two adjustments to retirement pay. The first is to adjust retirement pay to what it would have been under the 0.025-multiplier system. For example, a member retiring under REDUX with 20 YOS would receive 40 percent of *BP* between retirement and age 62 and 50 percent of *BP* thereafter. The second is to restore retirement pay to what it would have been had retirement pay been fully indexed to the CPI. Thus, retirement pay at age 62 is the same under both systems. After age 62, however, retirement pay under REDUX is once again adjusted according to the CPI minus one percent.

---

2 Calendar years of service are capped at 30 under all three systems.

APPENDIX B

The Active/Reserve Dynamic Retention Model

As discussed in Chapter Two, the model is a stochastic dynamic programming model of active retention and reserve participation at the individual level. The model is a theoretical basis for describing behavior where the individual is assumed to be rational and forward looking. In dynamic programming models, the current state depends on history, i.e., on the sequence of past states, and the decision taken in each period. The model is stochastic in the sense that in each period random factors enter the decision. There is a realization of the random factors in each period once it is reached. The realizations in future periods are not known in the current period, but the individual is assumed to know the distribution from which the random factors are drawn and can use this knowledge in developing an expected value of the future consequences of the current-period choice. The following expressions give the structure of the model. Again, the model is defined at the individual level, however, the individual subscript is suppressed for brevity.

$$Y_{jkt}(s_t, \varepsilon_{kt}; \gamma) = w_{kt} + \gamma_k + \beta \, \mathrm{Emax}\big(Y_{ka}(s_{t+1}, \varepsilon_{at+1}; \gamma), Y_{kr}(s_{t+1}, \varepsilon_{rt+1}; \gamma), Y_{kc}(s_{t+1}, \varepsilon_{ct+1}; \gamma)\big) + \varepsilon_{kt} \quad \text{(B.1)}$$

Y_{jk}= value function for transition from j to k, j, $k \in$ {active, reserve, civilian}
$s_t = (ay_t, ry_t, t)$ where ay = active years, ry = reserve years, t = total years
w_{kt}= current pay in k at t

$$\gamma_k = \begin{cases} \gamma_a & \text{monetary value of preference for serving in AC} \\ \gamma_r & \text{monetary value of preference for serving in RC} \\ 0 & \text{preference for civilian job} \end{cases}$$

β = personal discount factor
Emax = expected value of the maximum
ε_{kt} = random shock in k at t.

The model is generally structured to allow movement between active, reserve, and civilian statuses, but in applying the model we do not permit an individual who leaves an active component to reenter. This decreases the state space, which facilitates the estimation of the model, and reflects the fact that reentry is relatively rare. The value function Y_{jkt} subscripts indicate current status j, a status k that the individual can enter next period from j, and the time period. The value function is additively separable in the current pay in k, the monetary value of taste in k, the present value of being able to choose the best alternative in the following period given k in the current period, and the random shock in k in the current period.

The state, s_t, is defined in terms of active years, reserve years, and total years accumulated as of period t. Current pay depends on the state. Active pay is average annual regular military compensation given the number of active years and reserve years. Reserve pay is a fraction of reserve annual regular military compensation under the assumption that a reservist accumulates 75 points in a year and receives 75/360 of annual reserve RMC (Appendix A). Pay as a reservist, i.e., in the reserve status, is the sum of reserve pay and civilian pay, an approach that assumes reservists are also employed in civilian jobs. Civilian pay depends on total years. In addition, pay in the civilian status includes the present discounted value of the active or reserve military retirement benefit payment if the individual is eligible to receive it.

The term γ_k is the monetary value of the individual's preference relative to the civilian sector; i.e., γ_a for active service and γ_r for reserve service. The personal discount factor, β, is defined as $1/1 + r$, where r is the personal discount rate. The operator Emax gives the expected value of the *maximum* of the value functions in the next period. The Emax expression reflects the fact that the individual can reoptimize in the next period once the random shocks in that period have been realized. The current period assessment of the value of the best choice tomorrow is the expected value of the maximum of tomorrow's choices. The term ε_{kt} is the random shock in status k in period t.

The model is structured as a Markov process. In the next period there is a chance that any allowable status can be entered. Further, because the state is assumed to capture all relevant information from the individual's history and the random shocks are uncorrelated, it is possible to partition the expected value of the maximum given the current state. Using this insight, the model also can be written:

$$Y_{jkt}(s_t) = w_{kt} + \gamma_k + \beta \sum_m \pi_{km}(s_{t+1} \mid s_t) Y_{km}(s_{t+1}) + \varepsilon_{kt}$$

$$\pi_{km}(s_{t+1} \mid s_t) = \text{probability alternative } m \text{ is } \max(Y_{km}(s_{t+1})), m \in \{a, r, c\}. \quad \text{(B.2)}$$

The model assumes that a reservist holds a civilian job. This is a simplifying assumption because some reservists are full-time students, unemployed, or out of the labor force, but the idea is that participation in the reserves is concurrent with another main activity, a job. Therefore, a civilian job shock will be present in both the civilian and reserve statuses.

To allow for error correlation between the reserve and civilian alternatives, we use a nested logit form where these alternatives represent one nest and the active alternative is the other nest. The choice is between the active alternative and the better alternative in the reserve/civilian nest, i.e., the maximum of the reserve alternative and the civilian alternative. To shorten notation, we rewrite Equation B.1 as $Y_{kj}(s_t)=V_j+\varepsilon_j$, where V_j represents the non-stochastic terms on the right side, and the other arguments and time subscript have been omitted. Adapting the treatment of Ben-Akiva and Lerman (1985), we develop the nested logit specification from the following expressions:

$$\begin{gathered} V_a+\varepsilon_a \\ \max[V_r+\omega_r, V_c+\omega_c] + \upsilon_{rc}. \end{gathered} \tag{B.3}$$

The first expression in Equation B.3 corresponds to the active alternative, and the second expression corresponds to the reserve/civilian nest alternative. The active alternative can be thought of as a nest with a single element. The nested logit model assumes that ε_a has the same distribution as the sum of the errors in the second expression, so we need to ensure that this requirement is met. Also, we assume that all errors are generated from extreme value distributions. When the errors have the same extreme value distribution, and in particular have the same variance, then the choice between the nests has the logit form. Train (2003) provides a proof that when alternatives have identically distributed, independent extreme-value errors, the probability that a particular alternative is the maximum has a logit form. Ben-Akiva and Lerman (1985) show that the nested logit model can be written as a choice between alternatives, each of which is the maximum choice from its nest. As we show for our model, the errors of these maximum choices can be constructed to have the same variance; hence, Train's proof applies.

The extreme value distribution $EV[a,b]$ has the form $e^{-e^{(a-x)/b}}$ with mean $a+b\gamma$ and variance $\pi^2b^2/6$, where γ is Euler's Gamma ($\approx$0.577), a is the location parameter, and b is the scale parameter. The variance is proportional to the square of the scale parameter, and we use the fact that equal scale parameters imply equal variances. Let ω_r and ω_c in Equation B.3 be within-nest errors drawn from an extreme-value distribution $EV[0,\lambda]$ and let υ_{rc} be the nest-specific error for the reserve/civilian nest in Equation B.3, distributed as $EV[0,\tau]$. In other words, υ_{rc} can be thought of as a shock that affects both the reserve and the civilian alternatives, whereas ω_r and ω_c affect each alternative separately.

It is known that $\max[V_r+\omega_r, V_c+\omega_c]$ also follows an extreme-value distribution with the same scale as for ω_r and ω_c but a different mean, namely, $EV\left[\lambda\ln(e^{V_r/\lambda}+e^{V_c/\lambda}),\lambda\right]$. Notice that this mean is positive, assuming V_r and V_c are positive, whereas the distribution for ω_r and ω_c has a zero mean. Intuitively, the expected value of being able to choose the larger of two random draws, each with zero mean, is greater than zero. We rewrite the second expression in Equation B.3 as follows:

$$\lambda \ln(e^{V_r/\lambda} + e^{V_c/\lambda}) + \omega'_{rc} + \nu_{rc}, \text{ where} \tag{B.4}$$
$$\omega'_{rc} = \max\left[V_r + \omega_r, V_c + \omega_c\right] - \lambda \ln\left(e^{V_r/\lambda} + e^{V_c/\lambda}\right)$$
$$\omega'_{rc} \sim EV[0,\lambda].$$

Define $\varepsilon_{rc} = \omega'_{rc} + \upsilon_{rc}$. It is the sum of two independent, differently distributed extreme-value variables. The error ω'_{rc} is the single error associated with taking the maximum of V_r+ω_r and V_c+ω_c, and the definition of ω'_{rc} ensures that its mean is zero. Further, υ_{rc} is the single error at the nest level. The distributions of ω'_{rc} and υ_{rc} have the same location parameter (zero), but different scale parameters. In general, the variance of the sum of two independent random variables is the sum of the variances, so the variance of $\varepsilon_{rc} = \omega'_{rc} + \upsilon_{rc}$ is $\pi^2(\lambda^2 + \tau^2)/6$, implying a scale parameter for the R/C nest of $\sqrt{\lambda^2+\tau^2}$. It follows that $\varepsilon_{rc} \sim EV\left[0,\sqrt{\lambda^2+\tau^2}\right]$. We also want ε_a to have the same distribution (i.e., the same location and scale parameters), so we set $\varepsilon_a \sim EV\left[0,\sqrt{\lambda^2+\tau^2}\right]$. For brevity, let $\sqrt{\lambda^2+\tau^2}$.

Drawing this together, the model may be written as follows:

$$V_a + \varepsilon_a$$
$$\lambda \ln\left(e^{V_r/\lambda} + e^{V_c/\lambda}\right) + \varepsilon_{rc} \tag{B.5}$$
$$\varepsilon_a, \varepsilon_{rc} \sim EV\left[0,\kappa\right] .$$

Assuming that the individual chooses the higher-valued alternative, this leads to a probability of choosing *active* that has the logit form, as Train (2003) showed:

$$\Pr(active) = \frac{e^{V_a/\kappa}}{e^{V_a/\kappa} + e^{\lambda \ln\left(e^{V_r/\lambda} + e^{V_c/\lambda}\right)/\kappa}}$$
$$= \frac{e^{V_a/\kappa}}{e^{V_a/\kappa} + \left(e^{V_r/\lambda} + e^{V_c/\lambda}\right)^{\lambda/\kappa}} . \tag{B.6}$$

The second line follows from the fact that $e^{b \ln a} = a^b$.

The within-nest error terms, ω, are distributed $EV[0,\lambda]$ and the "total" error terms, ε, are distributed $EV\left[0,\sqrt{\lambda^2+\tau^2}\right]$.

Therefore, the fraction of the error variance accounted for by the within-nest, choice-specific, portion of the total error is

$$\frac{\lambda^2}{\tau^2 + \lambda^2} . \tag{B.7}$$

It follows that the fraction of the error variance attributable to the within-nest common shock is one minus this amount, or $\tau^2/(\tau^2 + \lambda^2)$.

As a thought experiment, we can think of the problem of selecting the best alternative from the nest as choosing between

$$
\begin{aligned}
&V_r + \omega_r + \upsilon_{rc} \\
&V_c + \omega_c + \upsilon_{rc}
\end{aligned}
\tag{B.8}
$$

The correlation between these two total utilities (viewed by themselves before one has been chosen) is

$$
\begin{aligned}
\rho &= \frac{Cov\left[V_r + \omega_r + \nu_{rc}, V_c + \omega_c + \nu_{rc}\right]}{\sqrt{Var\left[V_r + \omega_r + \nu_{rc}\right]}\sqrt{Var\left[V_c + \omega_c + \nu_{rc}\right]}} \\
&= \frac{\tau^2}{\tau^2 + \lambda^2} \; .
\end{aligned}
\tag{B.9}
$$

As shown in Equation B.9, a larger variance of the common shock results in a larger correlation between the reserve and civilian alternatives. Thus, the nested logit formulation succeeds in giving us a specification that allows the shocks to the reserve and civilian alternatives to be correlated, and the greater the common shock, the greater the correlation.

Applying the rule above for the distribution of the maximum of two values, we see that

$$
\begin{aligned}
&\max\left[V_a + \varepsilon_a, \lambda \ln\left(e^{V_r/\lambda} + e^{V_c/\lambda}\right) + \varepsilon_{rc}\right] \sim EV\left[\kappa \ln\left(e^{V_a/\kappa} + e^{\lambda \ln\left(e^{V_r/\lambda} + e^{V_c/\lambda}\right)/\kappa}\right), \kappa\right] \\
&EV\left[\kappa \ln\left(e^{V_a/\kappa} + e^{\lambda \ln\left(e^{V_r/\lambda} + e^{V_c/\lambda}\right)/\kappa}\right), \kappa\right] = EV\left[\kappa \ln\left(e^{V_a/\kappa} + \left(e^{V_r/\lambda} + e^{V_c/\lambda}\right)^{\lambda/\kappa}\right), \kappa\right].
\end{aligned}
\tag{B.10}
$$

As before, the second line follows from $e^{b \ln a} = a^b$.

Applying the formula for the mean of an extreme-value distribution to Equation B.10, the expected value of the maximum of the two alternatives (active versus the maximum of reserve/civilian), is

$$
\kappa\left(\gamma + \ln\left[e^{V_a/\kappa} + \left(e^{V_r/\lambda} + e^{V_c/\lambda}\right)^{\lambda/\kappa}\right]\right).
\tag{B.11}
$$

Further, given that active is not an option, the expected value of the maximum of the two alternatives (reserve and civilian) is

$$
\begin{aligned}
&\kappa\left(\gamma + \ln\left[\left(e^{V_r/\lambda} + e^{V_c/\lambda}\right)^{\lambda/\kappa}\right]\right) \\
&= \kappa\gamma + \lambda \ln\left[e^{V_r/\lambda} + e^{V_c/\lambda}\right].
\end{aligned}
\tag{B.12}
$$

The first line of Equation B.12 does not contain the term $e^{V_a/\kappa}$ because the constraint that the individual cannot reenter the active component means, in effect, that V_a is set to negative infinity, and $e^{-\infty} = 0$. The second line simplifies the log expression.

The expected value of the maximum of a set of choices is referred to as the surplus function, and the surplus function can be used to derive choice probabilities. The Williams-Daly-Zachary Theorem (see McFadden, 1981) states that the probability of choosing a given alternative equals the partial derivative of the surplus function with respect to the value of the alternative. Thus, the probability of choosing to remain in an active component is as follows:

$$\begin{aligned}\Pr(active) &= \frac{\partial\kappa\left(\gamma+\ln\left[e^{V_a/\kappa}+\left(e^{V_r/\lambda}+e^{V_c/\lambda}\right)^{\lambda/\kappa}\right]\right)}{\partial V_a} \\ &= \frac{e^{V_a/\kappa}}{e^{V_a/\kappa}+\left(e^{V_r/\lambda}+e^{V_c/\lambda}\right)^{\lambda/\kappa}}.\end{aligned} \tag{B.13}$$

This is the same as that shown in Equation B.6, which replicated the usual logit specification. To emphasize the meaning of Equation B.13, we restate it as

$$PR\left(V_a+\varepsilon_a>\lambda\ln\left(e^{V_r/\lambda}+e^{V_c/\lambda}\right)+\varepsilon_{rc}\right)=\frac{e^{V_a/\kappa}}{e^{V_a/\kappa}+\left(e^{V_r/\lambda}+e^{V_c/\lambda}\right)^{\lambda/\kappa}}.$$

By the same approach, the probabilities of choosing reserve and civilian are

$$\begin{aligned}\Pr(reserve) &= \frac{\left(e^{V_r/\lambda}+e^{V_c/\lambda}\right)^{\lambda/\kappa}}{e^{V_a/\kappa}+\left(e^{V_r/\lambda}+e^{V_c/\lambda}\right)^{\lambda/\kappa}}\frac{e^{V_r/\lambda}}{e^{V_r/\lambda}+e^{V_c/\lambda}} \\ \Pr(civilian) &= \frac{\left(e^{V_r/\lambda}+e^{V_c/\lambda}\right)^{\lambda/\kappa}}{e^{V_a/\kappa}+\left(e^{V_r/\lambda}+e^{V_c/\lambda}\right)^{\lambda/\kappa}}\frac{e^{V_c/\lambda}}{e^{V_r/\lambda}+e^{V_c/\lambda}}.\end{aligned} \tag{B.14}$$

Given that the individual has left the active component and cannot reenter it, the probabilities of choosing reserve or civilian are, respectively,

$$\begin{aligned}\Pr(reserve \mid inactive) &= \frac{\partial\left(\kappa\gamma+\lambda Log\left[e^{V_r/\lambda}+e^{V_c/\lambda}\right]\right)}{\partial V_r}=\frac{e^{V_r/\lambda}}{e^{V_r/\lambda}+e^{V_c/\lambda}} \\ \Pr(civilian \mid inactive) &= \frac{\partial\left(\kappa\gamma+\lambda Log\left[e^{V_r/\lambda}+e^{V_c/\lambda}\right]\right)}{\partial V_c}=\frac{e^{V_c/\lambda}}{e^{V_r/\lambda}+e^{V_c/\lambda}}.\end{aligned} \tag{B.15}$$

A comparison of Equations B.14 and B.15 shows that the probability of choosing to be a reservist equals the probability of choosing the reserve/civilian nest multiplied by the probability of choosing reserve, given that the individual is in the nest. A similar statement holds for the probability of choosing to be a civilian.

The model provides structure regarding the choice among alternatives. The choice is modeled to depend additively on current pay, taste for active or reserve service, current shock, and the expected value of rational choice among alternatives in the uncertain future, i.e., the expected value of the maximum. Further structure comes from assuming the individual knows the shock variances and so has the information needed to compute the expected value of the maximum. The individual also knows the civilian, reserve, and active pay lines. With this information, along with knowing the current state and shock draws, the individual can solve the problem and determine which alternative is best.

The position of the analyst is different. The analyst does not know the individual's tastes for active and reserve service, nor the current shocks, but is assumed to know their type of distribution. In particular, we assume that tastes follow a bivariate normal distribution and shocks follow generalized extreme value distributions, as assumed in the nested logit model. The bivariate normal distribution has five parameters: mean active taste, mean reserve taste, active taste variance, reserve taste variance, and active-reserve taste covariance or correlation. The extreme value distributions for the shocks have zero location parameters and non-zero scale parameters, hence non-zero variances.

In addition, we include the personal discount factor and switching cost parameters. The latter represent the cost of switching across states. There are switching costs for switching from active in the first two years of active service, switching from active in later years, and switching from civilian to reserve status. Estimation of the model involves finding the taste, shock, personal discount, and switching cost parameters that fit the data best, where the data consist of longitudinal observations on the individual's sequence of active, reserve, and civilian statuses over the work life.

Unlike the individual, who is assumed to have the information to make an explicit choice each period, the analyst does not have information about the individual's tastes or shocks. But the analyst can make use of the model and the functional form of the shock distributions to write an expression for the probability of a particular choice given the individual's state, and in this way can compose an expression for likelihood for the sequence of statuses over the individual's career. Still, this expression is conditional on the individual's tastes. Because these tastes are unknown to the analyst, they need to be integrated out of the expression, where the integration uses the assumption that the tastes have a bivariate normal distribution.

In estimation, the integration is done numerically. For each individual, a Halton sequence of 23 pairs of active and reserve seed tastes is drawn and then, using trial values of the taste distribution parameters, the Halton draws are translated as though they were drawn from the distribution of tastes given the trial values of the parameters. The translation is done via a Cholesky decomposition (Appendix C). For each resulting pair, the dynamic program is solved, giving values of the value functions at each decision point and hence values of the individual's career likelihood. The integration over tastes is accomplished by taking the average of the likelihoods over the 23 valuations.

The process of estimation tries different values of the parameters until the career likelihoods are maximized for the sample of service members used in the estimation. In many respects, this is standard maximum likelihood estimation, but it differs in two ways. First, the model has a specific structure for the value function, as mentioned above. Second, for each set of trial parameters, the dynamic programming problem must be re-solved for all periods for all 23 pairs of taste draws for each individual. Then, given the new solution of the model, the choice probabilities are updated, and the likelihood function is reevaluated to determine whether the fit has improved and in what direction the distribution parameters should be further changed in proceeding to the next iteration of estimation. Re-solving the dynamic program requires extensive computation. Estimating the Hessian matrix to determine the optimal direction of change is also computationally time consuming.

As the estimation algorithm iterates, we can think of the effect of changing the shock variances while holding constant the taste distribution parameters. An increase in a shock variance improves the fit if it does a better job of accounting for transitions from active to reserve, active to civilian, civilian to reserve, and reserve to civilian. That is, changing the variance affects all transitions, given any set of taste distribution parameters. Reasoning similarly, changing the mean active taste affects the fit of the active/active, active/reserve, and active/civilian transitions but not the other transitions. Changing the active taste variance helps account for dispersion in the transition probabilities from active duty. A change in the taste covariance affects the degree to which longer active careers are associated with longer reserve careers, e.g., higher transitions from civilian to reserve and lower transitions from reserve to civilians. Similar reasoning applies to the reserve taste mean and variance, with the implication that all the taste and shock parameters are identified. The personal discount rate is also estimable as it decreases future values relative to present values until the best fit is achieved. The switching cost parameters further improve the fit of the model.

We use the BHHH hill-climbing algorithm to optimize the likelihood function (Berndt et. al., 1974). We compute standard errors by numerically differentiating the likelihood function, evaluated at parameter estimates to produce a matrix of second derivatives, or Hessian matrix. The standard errors are the square root of the absolute value of the diagonal of the inverse of the Hessian.

APPENDIX C

Cholesky Decomposition and the Parameter Estimates

In estimating the dynamic program, the optimization routine requires iterative changes to the taste draws and the shock draws as the parameter estimates are updated. To implement these changes, the estimation routine expresses the draws as linear equations in variables that map back to the parameters to be estimated. This is accomplished by a Cholesky decomposition of the covariance matrix. Following Train (2003), we take two independent draws from a standard normal distribution ($N(0, 1)$) and use a Cholesky decomposition to transform them into random variables that are jointly normally distributed with means $\mu=(\mu_1, \mu_2)$ and covariance matrix

$$\Omega = \begin{pmatrix} \sigma_{11}^2 & \sigma_{12}^2 \\ \sigma_{21}^2 & \sigma_{22}^2 \end{pmatrix}.$$

The covariance matrix is positive definite and symmetric $\sigma_{12} = \sigma_{21} = \rho\sigma_{11}\sigma_{22}$, so we can define a Cholesky matrix

$$L = \begin{pmatrix} \alpha_{11} & 0 \\ \alpha_{21} & \alpha_{22} \end{pmatrix}$$

such that $LL' = \Omega$.

Let (η_1, η_2) be two draws from a standard normal and calculate $\varepsilon = \mu + L\eta$. The values $\varepsilon = (\varepsilon_1, \varepsilon_2)$ follow a bivariate normal with mean μ and covariance matrix Ω. The values are normally distributed because they are the sum of normals, which is normally distributed. As Train (2003) shows, the mean of ε is μ: $E(\varepsilon) = \mu + LE(\eta) = \mu$. The covariance of ε is Ω.

In our application,

$$\begin{pmatrix} \varepsilon_1 \\ \varepsilon_2 \end{pmatrix} = \begin{pmatrix} \mu_1 \\ \mu_2 \end{pmatrix} + \begin{pmatrix} \alpha_{11} & 0 \\ \alpha_{21} & \alpha_{22} \end{pmatrix} \begin{pmatrix} \eta_1 \\ \eta_2 \end{pmatrix} \quad \text{or} \quad \begin{aligned} \varepsilon_1 &= \mu_1 + \alpha_{11}\eta_1 \\ \varepsilon_2 &= \mu_2 + \alpha_{21}\eta_1 + \alpha_{22}\eta_2 \end{aligned}$$

So $Var(\varepsilon_1) = \alpha_{11}^2$, $Var(\varepsilon_2) = \alpha_{21}^2 + \alpha_{22}^2$ and $Cov(\varepsilon_1, \varepsilon_2) = \alpha_{11}\alpha_{21}$, which are the elements of the covariance matrix Ω.

As the estimation procedure iterates, the values of the parameters μ_1, μ_2, σ_{11}, σ_{22}, and ρ are updated, thereby shifting the means and changing the variance and covariance. These, in turn, cause the values of the tastes ε_1, ε_2 to change. As the estimation procedure iterates, the shape parameter of the shock distribution, τ, also changes. Given the changes in tastes and τ, each individual must re-solve his or her dynamic program at each state observed, and a new iteration begins. This continues until the best fit of the taste and shock parameters to the data is obtained.

Parameter Estimates

To make the numerical optimization easier, we do not estimate the parameters directly but instead estimate the natural logarithm of each parameter, with the exception of the correlation, for which we estimate the inverse hyperbolic tangent. To recover the parameters, we need to transform the estimates. The raw and transformed parameter estimates for each service, for officers and enlisted personnel, are shown in Tables C.1–C.4. All of the estimates are highly statistically significant. The transformed estimates in Tables C.2 and C.4 are denominated in thousands of dollars, except for the estimates of the taste correlation and the discount factor.

The model nests the "reserve" and "civilian" alternatives because most reservists also hold a civilian job, and a shock to "civilian" is therefore also likely to be felt by "reserve." At the same time, each alternative is likely to be subject to its own shock. This reasoning is confirmed by the results for tau and lambda, which are estimates for the nest shock and the alternative-specific shock, respectively. Both are both statistically significant. A shock variance is $\pi^2/6$ multiplied by the square of the shape parameter, so the standard deviation of a shock is $\pi / \sqrt{3} \approx 1.8$ multiplied by the shape parameter. The estimates of tau for enlisted personnel range from about \$35,000 to \$43,000, and the estimates of lambda are \$9,000 to \$12,000. A large value of tau compared to lambda means that a high fraction of the total error variance comes from the nest error, i.e., the error common to the reserve and civilian alternatives. If, as seems plausible, the nest error is mainly the result of random factors related to the civilian job, the results for both officers and enlisted personnel suggest that job-related shocks are an important source of randomness for individuals who served in each service.

The mean active taste estimates are approximately the same for the Navy and Marine Corps, at roughly –\$17,000, and higher for the Army and the Air Force at –\$10,000 to –\$9,000 for enlisted personnel. For officers, mean active taste ranges from \$0 to about –\$22,000. A generally negative mean active taste at entry is consistent with the military needing to set pay for enlistees entering the active component at a level considerably higher than the median civilian wage for individuals with equivalent education and experience. The higher military pay offsets the negative taste for military service. Note that these estimated taste means are for the population of enlistees who are at the start

Table C.1
Parameter Estimates and Standard Errors (SE) for Enlisted Personnel, by Service

	Army		Navy		Air Force		Marines	
Coefficient	**Estimate**	**SE**	**Estimate**	**SE**	**Estimate**	**SE**	**Estimate**	**SE**
ln(Tau)	3.763	0.050	3.738	0.064	3.564	0.067	3.620	0.085
ln(Lambda)	2.521	0.062	2.293	0.084	2.465	0.079	2.240	0.128
ln(-1*(Mean Active Taste))	2.229	0.042	2.752	0.049	2.319	0.050	2.911	0.063
ln(-1*(Mean Reserve Taste))	3.181	0.054	3.485	0.077	3.255	0.085	4.048	0.131
ln(SD Active Taste)	0.508	0.135	1.334	0.099	2.001	0.069	0.603	0.157
ln(SD Reserve Taste)	2.581	0.069	2.777	0.094	2.785	0.094	3.383	0.147
atanh(Taste Correlation)	0.789	0.018	1.117	0.033	0.708	0.025	1.294	0.028
ln(-1*(Leave Active in First Two Years)	1.657	0.166	3.914	0.066	4.028	0.067	4.229	0.086
ln(-1*(Switch Civilian to Reserve))	3.906	0.062	3.700	0.085	4.037	0.079	3.611	0.130
ln(-1*(Leave Active After First Two Years))	1.870	0.084	2.500	0.093	2.798	0.073	2.384	0.142
ln(Beta)	-0.131	0.004	-0.103	0.004	-0.129	0.005	-0.099	0.005
-1*Log Likelihood	125,434		93,697		101,264		80,369	
N	29,619		29,942		29,928		29,931	

NOTES: Tau is the shape parameter of the nest error; Lambda is the shape parameter of the error specific to each alternative in the nest (here, "reserve" and "civilian"); Leave Active in First Two Years is a switching cost; Switch Civilian to Reserve is a switching cost; Leave Active After First Two Years is a switching cost; Beta is the personal discount factor.

Table C.2
Transformed Parameter Estimates for Enlisted Personnel, by Service

Coefficient	Army	Navy	Air Force	Marines
Tau	43.059	42.026	35.303	37.327
Lambda	12.435	9.902	11.764	9.397
Mean Active Taste	–9.288	–15.677	–10.161	–18.368
Mean Reserve Taste	–24.061	–32.618	–25.923	–57.282
SD Active Taste	1.662	3.795	7.398	1.827
SD Reserve Taste	13.205	16.065	16.195	29.448
Taste Correlation	0.658	0.807	0.609	0.860
Leave Active in First Two Years	–5.241	–50.096	–56.135	–68.628
Switch Civilian to Reserve	–49.717	–40.435	–56.634	–36.993
Leave Active After First Two Years	–6.486	–12.183	–16.418	–10.845
Beta	0.878	0.902	0.879	0.906

NOTES: Transformed parameters are denominated in thousands of dollars, with the exception of Taste Correlation and Beta. Tau is the shape parameter of the nest error; Lambda is the shape parameter of the error specific to each alternative in the nest (here, "reserve" and "civilian"); Leave Active in First Two Years is a switching cost; Switch Civilian to Reserve is a switching cost; Leave Active After First Two Years is a switching cost; Beta is the personal discount factor.

of their military careers in the active component; as a cohort progresses the mean taste of the remaining population of enlistees will rise, not because individual tastes change, but because those individuals with lower taste are less likely to choose to reenlist.

The mean reserve taste estimates for enlisted personnel are in the neighborhood of –\$33,000 to –\$24,000 for the Army, Navy, and Air Force, but –\$57,000 for the Marine Corps. The lower values of the reserve taste compared to active taste indicate that many of those who served in an active component do not have nearly as strong an inclination to serve in a reserve component. This is borne out by the data that show only a minority of soldiers in any entering active cohort choosing to enter the selected reserve. The estimated standard deviation of active taste is about \$2,000 for the Marine Corps and Army, \$4,000 for the Navy, and \$7,000 for the Air Force. The estimates imply greater homogeneity of tastes among enlistees in the Marine Corps and Army, whereas tastes among Navy and Air Force enlistees are more diverse. Intuitively, a low variance in tastes means that, controlling for state, pay, and shocks, there will be little difference in active stay/leave decisions among individuals at a decision point. The estimated standard deviation of reserve taste is an order of magnitude higher. The estimates are \$13,000 for the Army, \$16,000 for the Navy and Air Force, and \$30,000 for the Marine Corps. Thus, not only are mean reserve tastes lower than mean active tastes, but tastes for the reserves vary much more widely than tastes for the actives.

Tastes for the actives and reserves are positively correlated. The taste correlation for Air Force and Army enlisted personnel is about 0.60 to 0.65 but higher for Navy and Marine Corps enlisted personnel at 0.80 to 0.86. For officers, the correlation is

Table C.3
Parameter Estimates and Standard Errors (SE) for Officers, by Service

	Army		Navy		Air Force		Marines	
Coefficient	Estimate	SE	Estimate	SE	Estimate	SE	Estimate	SE
ln(Tau)	5.128	0.086	5.077	0.097	4.961	0.259	5.075	0.167
ln(Lambda)	3.505	0.138	2.679	0.141	3.274	0.487	2.996	0.212
ln(-1*(Mean Active Taste))	-13.565	1482.921	3.093	0.155	2.651	0.693	-3.844	37.530
ln(-1*(Mean Reserve Taste))	4.600	0.124	4.312	0.125	4.620	0.405	4.129	0.186
ln(SD Active Taste)	1.011	0.952	2.263	0.259	3.278	0.454	-2.744	30.328
ln(SD Reserve Taste)	4.221	0.139	3.892	0.150	4.351	0.466	3.794	0.231
atanh(Taste Correlation)	0.776	0.029	0.630	0.021	0.856	0.083	0.530	0.064
ln(-1*(Leave Active in First 3-4 Years))	3.535	0.213	5.502	0.110	6.031	0.282	4.061	0.254
ln(-1*(Switch Civilian to Reserve))	4.356	0.144	3.339	0.152	4.082	0.493	3.446	0.226
ln(Beta)	-0.069	0.002	-0.060	0.003	-0.061	0.010	-0.071	0.004
-1*Log Likelihood	5.128	0.086	5.077	0.097	4.961	0.259	5.075	0.167
N	3,442		3,170		643		923	

NOTES: Tau is the shape parameter of the nest error; Lambda is the shape parameter of the error specific to each alternative in the nest (here, "reserve" and "civilian"); Leave Active in First Two Years is a switching cost; Switch Civilian to Reserve is a switching cost; Beta is the personal discount factor.

Table C.4
Transformed Parameter Estimates for Officers, by Service

Coefficient	Army	Navy	Air Force	Marines
Tau	168.649	160.288	142.707	159.990
Lambda	33.267	14.577	26.405	19.997
Mean Active Taste	0.000	–22.044	–14.167	-0.021
Mean Reserve Taste	-99.514	–74.576	–101.469	-62.119
SD Active Taste	2.749	9.613	26.529	0.064
SD Reserve Taste	68.117	49.020	77.545	44.438
Taste Correlation	0.650	0.558	0.694	0.485
Leave Active in First Two Years	–34.288	–245.117	–416.169	-58.027
Switch Civilian to Reserve	–77.917	–28.198	–59.258	-31.359
Beta	0.933	0.942	0.941	0.931

NOTES: Transformed parameters are denominated in thousands of dollars, with the exception of Taste Correlation and Beta. Tau is the shape parameter of the nest error; Lambda is the shape parameter of the error specific to each alternative in the nest (here, "reserve" and "civilian"); Leave Active in First Two Years is a switching cost; Switch Civilian to Reserve is a switching cost; Beta is the personal discount factor.

typically lower, ranging from 0.48 to 0.65. An implication of the positive correlation is that a high active-taste individual is likely to have longer retention in an AC, and after leaving is more likely to participate in an RC.

The estimated model contains two switching costs in the enlisted specification. These improve the fit of the model. "Leave active in first two years" reflects the implicit cost inhibiting an AC service member from leaving after the first or second year of service. Some individuals might prefer to leave, e.g., low-taste individuals who were induced to enter the military by a negative civilian shock, but the military, having invested in their recruiting and training, does not want them to leave until they have served in a unit. "Switch civilian to reserve" is the implicit cost of entering the reserves from the civilian status. The cost may be thought of as a monetary estimate of the cost of having less time and flexibility to take part in family and career pursuits if one joins the reserves. There is no explicit cost of joining the reserve, but the estimate indicates that people behave as though there is one. In addition, there is an implicit cost associated with switching from the active to the reserve after the first two years of service, which is captured by the "Leave active after first two years." The estimated cost of switching from active to reserve service is significantly lower than entering the reserves from civilian status, which may reflect efforts by the services to bring exiting active members directly into the reserve.

The personal discount factor estimates range from 0.88 to 0.91 for enlistees. A discount factor of 0.91 translates to a personal discount rate of 10 percent; a factor of 0.88 to 14 percent. The real federal discount rate is on the order of 3 percent, and the difference between a 10 percent or higher personal discount rate and the federal discount rate indicates the possibility of saving cost by converting deferred compensation to current compensation.

The model fits the data well as shown in Chapter Two.

APPENDIX D

Theory of Successful Reform

This appendix presents the more formal application of the Demange and Geoffard model (2006) to military retirement reform. It corresponds to the outline of the model in Chapter Six. Here we consider retirement reform in a general context, by not distinguishing between active or reserve retirement and by recognizing that retirement reform may be accompanied by a pay change, bonus, or a set of separation payments. Thus, the discussion focuses on "compensation reform" rather than "retirement reform."

The model notation is as follows:

t = member effort or career length
$m(t; a, b)$ = military compensation
$c(t)$ = value of outside opportunities
μ = taste for service
θ = random national security shock
$s(t; \mu, \theta)$ = defense output
$\lambda s(t; \mu, \theta)$ = taxpayer's value of defense output
$ks(t; \mu, \theta)$ = member's patriotic value of producing output
p = taxes paid by taxpayers = collected/spent by Congress.

In the expression for military compensation, a and b are parameters that define the structure of military compensation.

The model has three players: service members, taxpayers, and Congress. Each is assumed to maximize their welfare or value functions. Service members have heterogeneous tastes for military service (defined by μ) that are only observed by the members themselves, not by the taxpayers or Congress. The total number of service members is normalized to 1; the population of members is indexed by j ($j \in J$) uniformly distributed over $J = [0, 1]$ and the type of service member j is $\mu_j \in M$. The probability distribution of types of members is F and the density f is defined over a finite range $[\mu_l, \mu^u]$. The total number of taxpayers is also normalized to 1, for simplicity. The type of taxpayer is i ($i \in I = [0, 1]$). In the health care example of Demange and Geoffard, i is the patient and the random shock θ is a health shock specific to patient i and is indexed by i, as θ_i. In the compensation reform context, the national security shock affects defense output, and the shock would likely be common to all taxpayers. That is, there would only be one type of defense shock. The shock is assumed to have a probability distribution *G*, and is unknown to the taxpayer ex ante. Since taxpayers are identical ex ante, taxes are independent of the defense shock at the time taxes are paid.

Military compensation, outside opportunities,[1] and defense output depend on the service members' career length t. The parameters a and b in the military compensation function $m(t; a, b)$ are the policy variables that define the level and structure of military compensation. Taxpayers seek values of a and b that increase their utility, i.e., that enhance defense output or lessen their taxes. For example, suppose $m(t; a, b)$ were linear.

$$m(t; a, b) = b + at. \tag{D.1}$$

Here, the parameter b is a fixed component of military compensation that is independent of career length, such as a bonus, whereas the parameter a is the "incentive" for longer career lengths.

The first-best military compensation scheme is the one that maximizes the joint welfare of the three parties. Suppose $m^0(a^0, b^0)$ was the preexisting status quo compensation system that provided insufficient incentive for long careers. To induce longer careers, taxpayers would like to increase a. While service members would like $m(t)$ to be increased, longer careers mean higher taxes, p, for taxpayers. But longer careers may lead to higher service member productivity and greater national defense, so there may be a gain for taxpayers who are also consumers. As will be shown below, the first-best solution accounts for the joint value to taxpayers and service members. Congress is assumed to play a neutral role, raising and disbursing tax revenue.

If taste for service were perfectly observed, the first-best compensation system could be achieved and would involve choosing a pair (a, b) for each member that made the member better off or no worse off while improving the utility of the taxpayers. When taste is not observed, a single reform is chosen. The new scheme might be better for only a subset of members and might make some members worse off. The latter group would have reason to block the reform effort. Thus, given the effect of reform on the welfare of some members, the reform may not be politically feasible.

A politically feasible reform is more cost-effective from the standpoint of the taxpayers and makes a large enough proportion of service members better off so that service members do not want to block reform.

---

[1] In the Demange and Geoffard model, effort supply is the key decision variable. Here, we use career length because of its natural correspondence to the dynamic retention model and the military annuity formulas. That said, the model could be adapted to follow the original Demange and Geoffard formulation and focus on the service member effort decision. In such a formulation, the role of the promotion system and the structure of the pay table in embedding incentives for effort supply would be the key policy areas of interest. We do not focus on those issues here.

All parties are assumed to be risk neutral. Furthermore, it is assumed that the defense output function $s(t)$ has positive but diminishing returns. In addition, output depends on external factors, namely members' taste for service and shocks to national security. We assume that national security shocks are ordered such that higher shocks decrease output. An example of such a shock might be a cut in defense spending apart from personnel. A "low" shock might be a small-scale military contingency that requires the deployment of forces for a military operation; here, the low shock increases defense output. Stronger tastes for service also increase output. To summarize these assumptions, we have:

$$s_t(t;\, \mu, \theta) > 0,\ s_{tt}(t;\, \mu, \theta) < 0$$
$$s_\mu(t;\, \mu, \theta) > 0,\ s_\theta(t;\, \mu, \theta) < 0. \qquad \text{(D.2)}$$

We proceed as follows. First we write down the objective function or value function for each player and derive the conditions that the first-best and most efficient compensation scheme must meet. Next, we discuss how the inability to observe service members' tastes makes the first-best unattainable and, under a policy that applies the same parameters a and b to all, results in some members being made worse off. Thus, some members will be against reforms meant to improve efficiency. Next, we define the constraint that reforms must meet to be politically feasible and write down the conditions for solving for the optimal reform given the objectives of the triad and the constraints they face.

Objective Function for Each Player

The value function for service members is the sum of military compensation and the monetized patriotic value of providing defense minus the opportunity cost of staying in the military. Members are assumed to make decisions about career length after observing the defense shock θ.

$$V(t;\, \mu, \theta) = m(t) - c(t) + ks(t;\, \mu, \theta). \qquad \text{(D.3)}$$

If $m(t)$ is a linear function given in Equation D.1 and $c(t)$ is also linear and equal to $c(t)=wt$, then Equation D.3 can be written as:[2]

$$V(t;\, \mu, \theta) = b + at - wt + ks(t;\, \mu, \theta). \qquad \text{(D.4)}$$

[2] Alternatively, let $m_t = b_1 + at$ and $c_t = b_0 + wt$, so $m_t - c_t = (b_1 - b_0) + (a - w)t$ and define $b = b_1 - b_0$.

The individual's optimal career length, t^*, is characterized by the condition that the marginal value of staying in the military (in terms of compensation and patriotic value) equals the marginal opportunity cost.

$$\begin{aligned} m'(t) + ks'(t; \mu, \theta) &= c'(t) \\ a + ks'(t; \mu, \theta) &= w \\ a &= w - ks'(t; \mu, \theta). \end{aligned} \tag{D.5}$$

In this formulation, the marginal effects of an incrementally longer career are at a and w for the military and civilian wage, respectively, and these are assumed to be constant. In contrast, the marginal patriotic value depends on the effect of a longer career on defense output. This is assumed to increase at a decreasing rate. The third line of Condition D.5 implies that the incremental increase in military compensation from a longer career equals the incremental increase in civilian compensation minus the marginal patriotic value of the longer career. The latter depends on the marginal increase in defense output, which here is assumed to be positive throughout its range. Consequently, Condition D.5 indicates that the marginal increase in military compensation can be less than the marginal increase in civilian compensation: $w < a$. [If a and w varied, then as the marginal increase in defense output decreases (i.e., as $s' \rightarrow 0$, Condition D.5 requires that the military marginal effect approaches the civilian marginal effect ($a \rightarrow w$)].

The second line of Condition D.5 is shown graphically in Figure D.1, in the case where $m(t)$ and $c(t)$ are linear. The figure shows that an increase in the incentive parameter a from a^0 to a^1 increases the optimal career length from t^0 to t^1. This occurs because, to maintain equality in Condition D.5, an increase in a must be offset by a decrease in $ks'(t)$, which requires an increase in career length t.

The optimal career length t^* can be written in terms of the compensation parameter or $t^*(a)$. The variable t^* doesn't depend on b because b drops out of Condition D.5. The parameter b is determined by the participation constraint, defined below. Note that t^* also depends w, μ, and θ, i.e., $t^* = t^*(a; w, \mu, \theta)$.

Given $t^*(a)$, we can write the service member's indirect value function in terms of the compensation parameter a.

$$V(a, b; \mu, \theta) = b + (a - w)t^*(a; \mu, \theta) + ks(t^*(a); \mu, \theta).$$

The expected value of the indirect utility function is:

$$\begin{aligned} V(a,b;\mu) &= \int_{\tilde{\theta}} b + (a-w)t^*\left(a;\mu,\tilde{\theta}\right)d\tilde{\theta} + k\int_{\theta} s\left(t^*\left(a;\mu,\tilde{\theta}\right);\mu,\tilde{\theta}\right)d\tilde{\theta} \\ &= b + (a-w)E_{\theta}t^*\left(a;\mu,\tilde{\theta}\right) + kE_{\theta}s\left(t^*(a);\mu,\tilde{\theta}\right) \\ &= b + (a-w)T(a;\mu) + kS(a;\mu). \end{aligned} \tag{D.6}$$

Figure D.1
Optimal Career Length *t* for Service Member

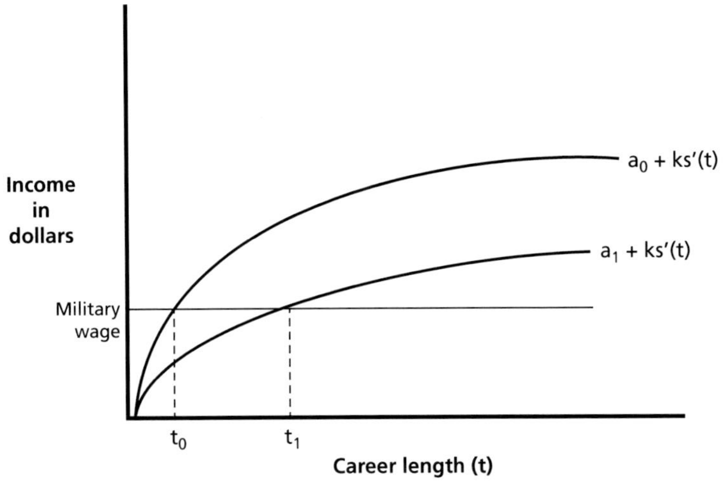

RAND *MG378-D.1*

In this expression, the random defense shock is denoted by θ. The term $S(a;\mu)$ is the expected value of defense output for a member with a given taste for service μ and $T(a,\mu)$ is the expected career length of a member with taste μ.

$$S(a;\mu) = E_\theta s\left(t^*\left(a;\mu,\tilde{\theta}\right);\mu,,\tilde{\theta}\right)$$
$$T(a;\mu) = E_\theta t^*\left(a;\mu,,\tilde{\theta}\right)$$

Turning next to the taxpayers/consumers, their utility or value function is the monetized value of national defense minus the taxes they pay. After the occurrence of the defense shock, the expected value for taxpayers, given that service member taste is unobserved, is:

$$U(a;\theta) = \lambda E_{\tilde{\mu}} s\left(t^*(a;\tilde{\mu},\theta);\tilde{\mu},\theta\right) - p. \qquad \text{(D.7)}$$

Ex ante, before the occurrence of the national security shock, expected utility is

$$U(a) = E_{\tilde{\theta}} U\left(a;\tilde{\theta}\right) = \int_{\tilde{\theta}} \lambda E_\mu s\left(t^*\left(a;\tilde{\mu},\tilde{\theta}\right);\tilde{\mu},\tilde{\theta}\right) d\tilde{\theta} - p = \lambda E_{\tilde{\mu}} S(a;\tilde{\mu}) - p. \qquad \text{(D.8)}$$

where from the definition above $S(a;\tilde{\mu}) = E_\theta s\left(t^*\left(a;\tilde{\mu},\tilde{\theta}\right);\tilde{\mu},\tilde{\theta}\right)$.

Here again, Congress is assumed to be a costless administrator that oversees the revenues collected from taxpayers and the expenditures of those revenues in the form of compensation to service members. Thus, the expected value function of Congress is tax revenues minus expected expenditures on military compensation:

$$V^{C}(a)=p-\int_{\theta} E_{\tilde{\mu}}m\left(t^{*}\left(a;\tilde{\mu},\tilde{\theta}\right)\right)d\tilde{\theta}=p-E_{\tilde{\mu}}m\left(T\left(a;\mu\right)\right)=p-b-aE_{\tilde{\mu}}T\left(a;\tilde{\mu}\right)\ . \qquad \text{(D.9)}$$

The term $E_{\tilde{\mu}}m\left(t^{*}\left(a;\tilde{\mu},\tilde{\theta}\right)\right)$ is expected military compensation where the expectation is taken over service member taste. When this expected value is integrated over national security shocks, it remains unchanged. This is the case because military compensation (see Equation D.1) is independent of the national security shock. To rationalize this, one might think of military compensation in Equation D.1 as being structured so as to fully insure service members against any risk (disutility) coming from national security shocks. This is a simplification; military compensation probably does include a compensating differential for ex ante risk related to national security shocks, but there are also deployment-related payments, health care and rehabilitative care, disability compensation, and survivor benefits realized ex post risk that are not covered in ex ante compensation.

First-Best Solution

The first-best military compensation structure is the one that leads to the career length that maximizes the expected joint welfare of the triad. That is, we find the optimal parameters a and b that lead to the career length that maximizes the sum of the expected value functions. Joint welfare is:

$$W(a)=\int_{\tilde{\theta}}\int_{\tilde{\mu}}\left[\lambda s(t)-p\right]+\left[p-m(t)\right]+\left[m(t)-c(t)+ks(t)\right]d\tilde{\mu}\,d\tilde{\theta} \qquad \text{(D.10)}$$

where $\lambda s(t) - p$ = taxpayer
$p - m(t)$ = Congress
$m(t) - c(t) + ks(t)$ = service member.

Consider the terms in the integrand. If tastes μ were known, then maximizing W would involve maximizing, for each individual, the difference

$$(\lambda + k)s(t(a;\mu,\theta)) - c(t), \text{ or}$$
$$(\lambda + k)s(t(a;\mu,\theta)) - wt(a;\mu,\theta).$$

The first-best career length, t^{fb} is the one that maximizes this difference. The first order condition is:

$$(\lambda + k)s'(t^{fb};\mu,\theta) = c'(t^{fb}) = w. \qquad \text{(D.11)}$$

The first-best career length equalizes the marginal value of experience in increasing defense output, accounting for both the value to taxpayers λ and to members in terms of patriotic value k to the opportunity cost of staying in the military. The first-best military compensation system has the parameter a^{fb} that leads to $t^{fb}(a^{fb})$. Further, we can use Equation D.11 to reexpress the optimal career length in Equation D.5 as a relationship for the first-best value of a. From Equation D.11, $(\lambda + k)s' = w$ so $s' = w/(\lambda + k)$. From Equation (5), $a = w - ks'$, so

$$a^{fb} = w - k\frac{w}{\lambda + k} = \frac{\lambda}{\lambda + k}w \tag{D.12}$$

This implies that the first-best incentive parameter a^{fb} has a single value for all service members regardless of their taste μ. Also note that a^{fb} is independent of b, the flat payment in the linear compensation scheme. The fact that a^{fb} is the same for all does not imply that the first-best career length t^{fb} is the same for all. The fact that it is not can be seen through an alternative expression for a^{fb} in terms of s'; from Equations D.5 and D.11 we have $a + ks' = w = (\lambda + k)s'$, and it follows that:

$$a^{fb} = \lambda s'(t^{fb}; \mu, \theta).$$

Tastes are not observed, so a first-best scheme that solved Equation D.12 for every value of μ could not be implemented. With tastes unobserved but with a known distribution, a^{fb} solves Equation D.11 for the average-taste member; most members will have values of μ where this condition does not hold exactly. Some service members may be made worse off by moving to the first-best compensation system from the status quo system. Although the new system is the best from the standpoint of maximizing expected joint welfare, it is not the first-best from the standpoint of maximizing every individual service member's welfare. These individuals may want to block reform efforts to move compensation to the first-best expected welfare solution. Thus, there is a political constraint to reaching the first-best solution. The question is how to improve on the status quo system when some members may block reform because the change will make them worse off. Put differently, how can military compensation reform be designed to be politically feasible?

The Political Feasibility Constraint

For a compensation reform to be politically feasible, it must be the case that at least q percent of service members prefer the new compensation system $m(t; a, b)$ over the status quo system $m(t; a^0, b^0)$. More formally, politically feasible compensation systems with parameters (a, b) satisfy

$$F[\mu \mid V(a, b; \mu) \geq V(a^0, b^0; \mu)] \geq q \tag{D.13}$$

where $F[\mu]$ is the cumulative distribution function of tastes for service. Recall from Equation D.6 that $V(a, b; \mu)$ is the expected value of military service to the individual with taste μ, where the expectation is taken over national security shocks. It is for this reason that θ, representing a security shock, does not appear in $V(a, b; \mu)$. Equation D.13 represents a political constraint that any new compensation reform must meet in addition to the "participation constraint" it must meet.

The Participation Constraint

The participation constraint is the condition requiring that the new compensation system sets pay sufficiently high to attract service members to the military, i.e., it induces them to choose to participate. If $\underline{\nu}$ is the reservation value for the service member with taste μ, the participation constraint is

$$V(a,b;\mu) \geq \underline{\nu}. \tag{D.14}$$

Balanced-Budget Constraint

A final constraint that must be met is that tax collections equal expected expenditures on military compensation. For simplicity, we assume Congress fulfills both roles at zero cost. The constraint is:

$$p = E_{\tilde{\theta}\tilde{\mu}} m\left(t\left(a;\tilde{\mu},\tilde{\theta}\right)\right) = b + aE_{\tilde{\mu}}T\left(a;\tilde{\mu}\right). \tag{D.15}$$

Achieving a Politically Feasible Reform

We now consider the conditions that lead to the optimal compensation scheme, defined by the parameters a and b, given the political feasibility constraint, the participation constraint, and the balanced-budget constraint. The optimal compensation system is the one that maximizes the taxpayer value function subject to the three constraints.

The taxpayer value function is given by Equation D.8. Internalizing the budget constraint of Equation D.15 into Equation D.8, we get:

$$U(a) = \lambda E_{\tilde{\mu}}S(a;\tilde{\mu}) - p = \lambda E_{\tilde{\mu}}S(a;\tilde{\mu}) - b - aE_{\tilde{\mu}}T(a;\tilde{\mu}). \tag{D.16}$$

The optimal reform $m(T(a))$, given the constraints, solves the problem:

$$\max_{a} U(a) = \lambda E_{\tilde{\mu}} S(a;\tilde{\mu}) - a E_{\tilde{\mu}} T(a;\tilde{\mu})$$

subject to

$$V\left(a,b;\mu^{c}(a,b)\right) \geq V\left(a^{0},b^{0};\mu^{c}(a,b)\right)$$
$$V(a,b;\mu) \geq \underline{\nu}. \quad \text{(D.17)}$$

where $\mu^c(a, b)$ is the taste for service of the service member with the "pivotal" taste. For political feasibility, a parameter pair (a, b) must be such that the fraction of members whose expected value of military service has increased is at least equal to q. As a and b increase, the expected value of military service increases, and as a consequence the fraction of members willing to support the reform also increases. At some point, that fraction will reach the level needed for political feasibility. One can define a set of pairs (a, b) for which the political feasibility constraint holds with equality: $\{a, b \mid F[\mu \mid V(a, b; \mu) \geq V(a^0, b^0; \mu)] = q\}$. For a pair (a,b) in the set, the taste at which the political feasibility condition holds defines the pivotal taste $\mu^c(a, b)$ for that pair.[3] The issue is to find the optimal pair in the set. This will be the value of a that maximizes taxpayer value and the value of b that satisfies the political feasibility constraint.

The second constraint in Equation D.17 is the participation constraint. We assume that if the pivotal member participated under the status quo, he will participate under the reform as well. Therefore, if the first constraint is satisfied, then so is the second.

Following Demange and Geoffard, it is useful to consider the problem from the perspective of a change from status quo. That is, we will consider $U(a, b) - U(a^0, b^0)$, the change in utility for the taxpayer/consumer when a and b are changed. As Demange and Geoffard note, "whenever the status quo differs from the first-best scheme ($a \neq a^{fb}$) welfare W can be increased." The intuition behind considering a change from status quo is to place the focus on the importance of attracting the pivotal member. The conditions for doing so, shown below, trade off the gain from moving a toward a^{fb} versus the cost of paying enough to attract the pivotal member relative to the average cost that would have been paid under the new system if the political feasibility constraint did not have to be met.

The change $U(a, b) - U(a^0, b^0)$ equals the change in welfare $W(a, b)$ and the change in the expected welfare of service members:[4]

$$U(a,b) - U\left(a^{0},b^{0}\right) = \left[W(a,b) - W\left(a^{0},b^{0}\right)\right] - E_{\tilde{\mu}}\left[V(a,b;\tilde{\mu}) - V\left(a^{0},b^{0};\tilde{\mu}\right)\right]. \quad \text{(D.18)}$$

---

3 See Demange and Geoffard (2006) for further discussion of the pivotal member.

4 Recall from inspection of Equation D.10 that Congress drops out of the welfare expression, so the change in welfare equals the change in taxpayer utility plus the change in the value of service to the service member. Equation D.18 places the change in taxpayer utility on the left-hand side and the change in welfare on the right-hand side.

Consider the second term on the right side of Equation D.18. Define $b(a, \mu)$ as the flat payment that would make a service member with taste μ indifferent between the old scheme and the new scheme. That is, define $b(a,\mu)$ such that the following expression is true:

$$V(a, b; \mu) - V(a^0, b^0; \mu) =$$
$$b(a, \mu) + (a - w)T(a; \mu) + kS(a; \mu) - (b^0 + (a^0 - w)(T(a^0; \mu) + kS(a^0; \mu)) = 0.$$

But because μ is unknown, the flat payment may be set at a value of b that is different than $b(a, \mu)$. Since the optimal career length and the optimal value of a were independent of b and the compensation scheme is linear, the flat payment can be changed without changing optimal career length and a. This implies that

$$V(a, b; \mu) - V(a^0,b^0;\mu) = b - b(a, \mu). \qquad \text{(D.19)}$$

If $b = b(a, \mu)$, the left-hand side holds as an equality, as in the line above, and if, say, $b > b(a, \mu)$, then $V(a, b; \mu) > V(a^0, b^0; \mu)$. The flat payment $b(a, \mu)$ can be thought of as the compensating variation associated with changing a from the status quo. If taste was known, it would be possible to set the flat payment separately for each member so that $b = b(a, \mu)$ and the member would be as well off in terms of expected utility under the new policy a as under the status quo a^0. However, μ is unknown, so it is it not possible to make the flat payment contingent on μ and a. Thus, for some members, b and $b(a, \mu)$ will differ. The difference will be the difference in expected utility, as shown in Equation D.19; the quantity $b - b(a, \mu)$ is the "information rent" received by the individual. Equation D.19 establishes that the information rent to a member with taste μ equals the difference in the value of military service under the new policy less that of under the old policy.

In meeting the political feasibility constraint, the flat payment b will be set such that the pivotal member's expected value of service under the new compensation scheme equals its value under the status quo: $b = b(a, \mu^c)$. That is, when the flat payment cannot be made contingent on taste, changing a from the status quo, a^0, requires that b be adjusted at $b(a,\mu^c)$ so as to "buy" the support of the pivotal member. This value $b(a, \mu^c)$ is paid to all service members, whereas if each member's taste were known the payment would be based on individual taste: $b(a, \mu)$. Having to pay $b(a, \mu^c)$ to all under the new scheme imposes an informational cost. The expected value of this cost per member is:

$$C(a,\mu^c) = E_{\tilde{\mu}}\left[V(a,b;\tilde{\mu}) - V(a^0,b^0;\tilde{\mu})\right] = b(a,\mu^c) - E_{\tilde{\mu}} b(a,\tilde{\mu}). \qquad \text{(D.20)}$$

Plugging $C(a, \mu^c)$ into Equation D.18 above:

$$U(a, b) - U(a^0, b^0) = [W(a, b) - W(a^0, b^0)] - C(a, \mu^c). \qquad \text{(D.21)}$$

The change in the utility of taxpayers equals the change in overall welfare minus the informational cost.

Reform involves changing the parameter a, and this change causes two types of effects. The first occurs because career length T changes conditionally on a. As shown above, career length depends on a and μ but not on the flat payment b. The second type of effect occurs because the flat payment is set high enough to attract the pivotal member; changing the flat payment does not change the optimal career length but will change the information cost. The marginal effects of a change in a on welfare is:

$$\begin{aligned} Wa &= \frac{d\left((\lambda+k)E_{\tilde{\mu}}S\left(T(a;\tilde{\mu})\right)-wE_{\tilde{\mu}}T(a;\tilde{\mu})\right)}{da} \\ &= \left((\lambda+k)E_{\tilde{\mu}}S'(T)-w\right)\frac{dE_{\tilde{\mu}}T(a;\tilde{\mu})}{da} \\ &= \frac{\lambda+k}{k}\left(a^{fb}-a\right)\frac{dE_{\tilde{\mu}}T(a;\tilde{\mu})}{da}. \end{aligned}$$

This result implies that the marginal increase in welfare is greater the farther a is from its first-best value and the greater the responsiveness of expected career length across service members to an incremental increase in a. In other words, if the system is initially far from the first-best, the gain to moving closer to the first-best is large, but if the system is close to the first-best, the gain is small.

For the marginal effect of a on informational cost, consider the terms to the right of the equal sign in Equation D.20. With respect to the first of these terms, observe that since $b(a, \mu^c) + (a - w)T(a, \mu^c) + S(a, \mu^c) = b^0 + (a^0 - w)T(a^0, \mu^c) + S(a^0, \mu^c)$ we have

$$b(a, \mu^c) = -\left((a - w)\,T(a, \mu^c) + S(a, \mu^c)\right) + b^0 + (a^0 - w)\,T(a^0, \mu^c) + S(a^0, \mu^c)$$

Recognizing that career length does not change as b changes, we get:

$$\frac{db(a,\mu^c)}{da} = -T(a,\mu^c).$$

The second term on the right-hand side of Equation D.20 is for non-pivotal members. For a non-pivotal member under the new policy, we use the same approach as above and write

$$b(a,\mu) = -\left((a - w)\,T(a, \mu) + S(a, \mu)\right) + b^0 + (a^0 - w)\,T(a^0, \mu) + S(a^0, \mu).$$

The marginal effect of a non-contingent change in a is $\frac{db(a,\mu)}{da} = -T(a,\mu)$, and the expected value of this change across service members is:

$$\frac{db(a)}{da} = -E_{\mu}T(a,\mu).$$

Recall that information cost is the amount that buys the support of the pivotal member to accept the reform, or $C(a,\mu^c)=b(a,\mu^c)-E_{\tilde{\mu}}b(a,\tilde{\mu})$. The marginal effect of a (non-contingent) change in a on information cost is:

$$C_a = E_{\tilde{\mu}}T(a,\mu)-T(a,\mu^c).$$

The information cost increases as the difference grows greater in the flat payment for the pivotal member versus what the average flat payment under the new policy would have been in the absence of the political feasibility constraint. Adding to this insight, we see that the marginal effect of a change in a on information cost depends on the difference between what average career length across members would have been under the new policy without the political feasibility constraint, and career length under the new policy for the pivotal member. That is, as the reform increases the incentive for longer careers, a, the increase in the information cost grows as the difference increases between the career length of the average member and the career length of the pivotal member.

Menu of Reform Plans

At the heart of political infeasibility of reform is the heterogeneity of service members' taste for service. Changing the compensation system to increase incentives for longer careers has a positive effect on career lengths and defense output, but it also has an opportunity cost, $c(t) = wt$, and increases taxes. For the new system to be politically feasible, it must be preferred by a critical mass of members, defined by the member with the pivotal taste for service. The value pivotal taste μ^c is the value of μ such that the proportion of members who serve longer than μ^c is q. However, to prevent q members from blocking reform, a compensating differential of b must be paid to all service members, including those who do not support the changes, to induce them to accept reform, as shown in Equation D.21. However, it may be the case that the payment b is so high it outweighs the welfare gain $W(a, b)$ relative to the status quo $W(a^0, b^0)$ in Equation D.21. That is, in Equation D.20, we get:

$$U(a, b) - U(a^0, b^0) = [W(a, b) - W(a^0, b^0)] - [b(a, \mu^c) - E_{\mu}b(a, \mu)] < 0. \quad \text{(D.22)}$$

The "gain" associated with reform is negative for taxpayers because the buyoff to members must be so large that it is no longer worthwhile to pursue the effort. The payment must be high because some members have a very low taste for service (assuming taste and career length are complements).

A menu approach provides an opportunity to tailor reform options to heterogeneous tastes and can induce sorting of members to reform options. Specifically, offering a menu that includes the status quo improves the likelihood of reform. Those with a low taste for service choose the existing system. Thus, those most likely to block reform in the absence of a menu are no worse off under a reform option that includes a menu. Those members better off under reform enjoy the gains to their value functions, as do taxpayers.

To see how a menu improves the likelihood of reform, define (a^1, b^1) as the compensation reform when no menu is offered and $(a_m^{\ 1}, b_m^{\ 1})$ as the compensation reform when a menu is offered. In both cases, the status quo is (a^0, b^0). When no menu is offered, the entire force G chooses the (a^1, b^1) reform option (since no other choice is offered). When the menu is offered, G_1 is the subset of members who choose $(a_m^{\ 1}, b_m^{\ 1})$, and G_0 is the subset who remains with the status quo (a^0,b^0) where $G_0+G_1= G$.

It is clear that $b_m^{\ 1} < b^1$. Those most likely to block reform and with the lowest taste for service μ will choose the status quo (a^0, b^0) and will not switch to the new plan. Only the most eager will choose the new plan $(a_m^{\ 1}, b_m^{\ 1})$. Thus, the compensating differential (flat payment) required to induce sufficient support for the new reform is lower. A smaller payment is needed to buy support for the reform because those who are least likely to support the reform are permitted to stay under the old system. In the absence of a menu, Equation D.21 becomes:

$$\Delta U = \left[W\left(a^1,b^1\right) - W\left(a^0,b^0\right)\right] - \left[E_{\tilde{\mu}}V\left(a^1,b^1;\tilde{\mu}\right) - E_{\tilde{\mu}}V\left(a^0,b^0;\tilde{\mu}\right)\right] \quad \textit{for } G$$

while with a menu, Equation D.21 becomes:

$$\Delta U_m^1 = \left[W\left(a_m^1,b_m^1\right) - W\left(a^0,b^0\right)\right] - \left[E_{\tilde{\mu}}V\left(a_m^1,b_m^1;\tilde{\mu}\right) - E_{\tilde{\mu}}V\left(a^0,b^0;\tilde{\mu}\right)\right] \quad \textit{for } G^1$$

Because $b_m^{\ 1} < b^1$, the change in the second brackets is smaller when there is a menu. Consequently, the change in taxpayer utility, ΔUm, is larger. Therefore, the likelihood that reform is politically feasible is greater.

Finally, as discussed in Demange and Geoffard, the menu of reforms could cross-subsidize each other. If the constraint is only that the sum of tax revenues equals the total outlays, it is possible that one reform subsidizes another. This feature can increase the feasible set of reforms.

Bibliography

Arkes, Jeremy, and M. Rebecca Kilburn, *Modeling Reserve Recruiting: Estimates of Enlistments*, Santa Monica, Calif.: RAND Corporation, MG-202-OSD, 2005. As of September 28, 2012: http://www.rand.org/pubs/monographs/MG202.html

Asch, Beth, and James Hosek, *Military Compensation: Trends and Policy Options*, Santa Monica, Calif.: RAND Corporation, DB-273-OSD, 1999. As of September 28, 2012: http://www.rand.org/pubs/research_briefs/RB7533.html

———, *Looking to the Future: What Does Transformation Mean for Military Manpower and Personnel Policy?* Santa Monica, Calif.: RAND Corporation, OP-108-OSD, 2004. As of October 17, 2012: http://www.rand.org/pubs/occasional_papers/OP108.html

———, "An Assessment of Recent Congressional Proposals," *The New Guard and Reserve*, Falcon Books, 2008, pp. 187–196.

Asch, Beth J., James Hosek, and David S. Loughran, *Reserve Retirement Reform: A Viewpoint on Recent Congressional Proposals*, Santa Monica, Calif.: RAND Corporation, TR-199-OSD, 2006. As of September 28, 2012:
http://www.rand.org/pubs/technical_reports/TR199.html

Asch, Beth J., Richard Johnson, and John T. Warner, *Reforming the Military Retirement System*, MR-809-OSD, Santa Monica, Calif.: RAND Corporation, 1998. As of September 28, 2012: http://www.rand.org/pubs/monograph_reports/MR748.html

Asch, Beth J., and John T. Warner, *A Theory of Military Compensation and Personnel Policy*, Santa Monica, Calif.: RAND Corporation, s-439-OSD, 1994a. As of September 28, 2012: http://www.rand.org/pubs/monograph_reports/MR439.html

———, *A Policy Analysis of Alternative Military Retirement Systems*, Santa Monica, Calif.: RAND Corporation, MR-645-OSD, 1994b. As of September 28, 2012: http://www.rand.org/pubs/monograph_reports/MR465.html

———, *Separation and Retirement Incentives in the Federal Civil Service: A Comparison of the Federal Employees Retirement System and Civil Service Retirement System*, Santa Monica, Calif.: RAND Corporation, MR-986-OSD, 1999. As of September 28, 2012: http://www.rand.org/pubs/monograph_reports/MR986.html

———, *The Thrift Savings Plan: Will Reservists Participate?* Santa Monica, Calif.: RAND Corporation, DB-306-OSD, 2000. As of September 28, 2012: http://www.rand.org/pubs/documented_briefings/DB306.html

———, "A Theory of Compensation and Personnel Policy in Hierarchical Organizations with Application to the United States Military," *Journal of Labor Economics*, Vol. 19, No. 3, July 2001, pp. 523–562.

Ben-Akiva, Moshe, and Steven Lerman, *Discrete Choice Analysis: Theory and Application to Travel Demand*, Cambridge, Mass.: MIT Press, 1985.

Berndt, Ernst, Bronwyn Hall, Robert Hall, and Jerry Hausman, "Estimation and Inference in Nonlinear Structural Models," *Annals of Economic and Social Measurement*, Vol 3/4, 1974, pp. 653–665.

Berkovec, James, and Steven Stern, "Job Exit Behavior of Older Men," *Econometrica*, Vol. 59, No. 1, 1991, pp. 189–210.

Christian, John, *An Overview of Past Proposals to Military Retirement Reform*, Santa Monica, Calif.: RAND Corporation, TR-376-OSD, 2006. As of September 28, 2012: http://www.rand.org/pubs/technical_reports/TR376.html

Coffey, Kenneth, Associate Director, National Security and International Affairs Division, General Accounting Office, testimony before the House Armed Services Subcommittee on Military Personnel and Compensation, July 29, 1983, 98th Congress.

Daula, Thomas, and Robert Moffitt, "Estimating a Dynamic Programming Model of Army Reenlistment Behavior," *Military Compensation and Personnel Retention: Models and Evidence*, Alexandria, Va.: U.S. Army Research Institute for the Behavioral and Social Sciences, February 1991, pp. 181–201.

Defense Advisory Committee on Military Compensation, *The Military Compensation System: Completing the Transition to an All-Volunteer Force*, April 2006.

Defense Business Board, "Modernizing the Military Retirement System: Task Group," unpublished draft briefing, July 21, 2011.

Defense Manpower Commission, *Defense Manpower: The Keystone to National Security, Report to the President and the Congress*, Washington D.C.: U.S. Government Printing Office, April 1976.

Demange, Gabrielle, and Pierre Yves Geoffard, "Reforming Incentive Schemes Under Political Constraint: The Physician Agency," *Annals of Economics and Statistics*, No. 83/84, 2006, pp. 221–250.

DoD — *See* U.S. Department of Defense.

Employee Benefit Research Institute, "Employment-Based Retirement Plan Participation: Geographic Differences and Trends (2005)," *Issue Brief No. 299*, Washington, D.C., 2006.

Freedberg, Sydney J., "Retirement Redux," *GovExec.com*, April 1, 1999. As of September 28, 2012: http://www.govexec.com/magazine/1999/04/retirement-redux/5994/

Fricker, Ronald D., *The Effects of Perstempo on Officer Retention in the U.S. Military*, Santa Monica, Calif.: RAND Corporation, MR-1556-OSD, 2002. As of September 28, 2012: http://www.rand.org/pubs/monograph_reports/MR1556.html

Goldberg, Matthew, *A Survey of Enlisted Retention: Models and Findings*, Alexandria, Va.: Center for Naval Analyses, 2001.

Gotz, Glenn A., and John McCall, *A Dynamic Retention Model for Air Force Officers: Theory and Estimates*, Santa Monica, Calif.: RAND Corporation, R-3028-AF, 1984. As of September 28, 2012: http://www.rand.org/pubs/reports/R3028.html

Gustman, Alan, Olivia Mitchell, and Thomas Steinmeier, "The Role of Pensions in the Labor Market: A Survey of the Literature," *Industrial Labor Relations Review*, Vol. 47, No. 3, 1994, pp. 17–438.

Hansen, Michael, and Jennifer Wenger, "Is Pay Responsiveness of Enlisted Personnel Decreasing?" *Defense and Peace Economics*, Vol. 16, 2005, pp. 29–43.

Heckman, James J., "Statistical Models for Discrete Panel Data," *Structural Analysis of Discrete Data and Econometric Applications*, Cambridge, Mass.: MIT Press, 1981, pp. 179–197.

Hook Commission, *Career Compensation for the Uniformed Services*, Washington D.C.: U.S. Government Printing Office, December 1948.

Hosek, James, and Michael G. Mattock, *Learning About Quality: How the Quality of Military Personnel Is Revealed Over Time*, Santa Monica, Calif.: RAND Corporation, MR-1593-OSD, 2003. As of September 28, 2012:
http://www.rand.org/pubs/monograph_reports/MR1593.html

Hosek, James, Beth G. Asch, and Michael G. Mattock, *Should the Increase in Military Pay Be Slowed?* Santa Monica, Calif.: RAND Corporation, TR-1185-OSD, 2012. As of October 17, 2012:
http://www.rand.org/pubs/technical_reports/TR1185.html

Hosek, James, Michael G. Mattock, C. Christine Fair, Jennifer Kavanagh, Jennifer Sharp, and Mark E. Totten, *Attracting the Best: How the Military Competes for Information Technology*, Santa Monica, Calif.: RAND Corporation, MG-108-OSD, 2004. As of September 28, 2012:
http://www.rand.org/pubs/monographs/MG108.html

Hosek, James, and Mark E. Totten, *Does Perstempo Hurt Reenlistment: The Effect of Long or Hostile Perstempo on Reenlistment*, Santa Monica, Calif.: RAND Corporation, MR-990-OSD, 1998. As of September 28, 2012:
http://www.rand.org/pubs/monograph_reports/MR990.html

———, *Serving Away from Home: How Deployments Influence Reenlistment*, Santa Monica, Calif.: RAND Corporation, MR-1594-OSD, 2002. As of September 28, 2012:
http://www.rand.org/pubs/monograph_reports/MR1594.html

Kavanagh, Jennifer, *Determinants of Productivity for Military Personnel: A Review of Findings on the Contribution of Experience, Training, and Aptitude to Military Performance*, Santa Monica, Calif.: RAND Corporation, TR-193-OSD, 2005. As of September 28, 2012:
http://www.rand.org/pubs/technical_reports/TR193.html

Keane, Michael, and Kenneth Wolpin, "The Solution and Estimation of Discrete Choice Dynamic Programming Models by Simulation and Interpolation: Monte Carlo Evidence," *Review of Economics and Statistics*, Vol. 76, 1994, pp. 648–572.

Laibson, David, "Golden Eggs and Hyperbolic Discounting," *Quarterly Journal of Economics*, Vol. 112, No. 2, May 1997, pp. 443–477.

Loughran, David S., Jacob Alex Klerman, and Craig Martin, *Activation and the Earnings of Reservists*, Santa Monica, Calif.: RAND Corporation, MG-474-OSD, 2006. As of September 28, 2012:
http://www.rand.org/pubs/monographs/MG474.html

Loughran, David S., Jacob Alex Klerman, and Bogdan Savych, *The Effect of Reserve Activations and Active-Duty Deployments on Local Employment During the Global War on Terrorism*, Santa Monica, Calif.: RAND Corporation, TR-321-RC, 2006. As of September 28, 2012:
http://www.rand.org/pubs/technical_reports/TR321.html

Mattock, Michael G., and Jeremy Arkes, *The Dynamic Retention Model for Air Force Officers: New Estimates and Policy Simulations of the Aviator Continuation Pay Program*, Santa Monica, Calif.: RAND Corporation, TR-470-AF, 2007. As of September 28, 2012:
http://www.rand.org/pubs/technical_reports/TR470.html

Mattock, Michael G., James Hosek, and Beth J. Asch, *Reserve Participation and Cost Under a New Approach to Reserve Compensation*, Santa Monica, Calif.: RAND, MG-1153-OSD, 2012. As of September 28, 2012:
http://www.rand.org/pubs/monographs/MG1153.html

McFadden, Daniel L., "Econometric Models of Probabilistic Choice," *Structural Analysis of Discrete Data with Econometric Applications*, Cambridge, Mass.: MIT Press, 1981, pp. 198–272.

Milgrom, Paul, "Employment Contracts, Influence Activities, and Efficient Organization Design," *Journal of Political Economy*, Vol. 96, No. 1, 1988, pp. 42–60.

Office of the Assistant Secretary of Defense, Reserve Affairs, *2007 Official Guard and Reserve Manpower Strengths and Statistics Report*, undated.

———, *2009 Official Guard and Reserve Manpower Strengths and Statistics Report*, undated.

Office of the Under Secretary of Defense, *National Defense Budget Estimates for FY 2007 (Green Book)*, 2006.

Papke, Leslie, E. "Are 401(k) Plans Replacing Other Employer-Provided Pensions: Evidence from Panel Data," *Journal of Human Resources*, Vol. 34, No. 2, Spring 1995, pp. 311–325.

———, "Participation in and Contributions to 401(k) Pension Plans: Evidence from Plan Data," *Journal of Human Resources*, Vol. 30, No. 2, Spring 1999, pp. 346–368.

Poterba, James, Steven Venti, and David Wise, "401(k) Plans and Future Patterns of Retirement Savings," *American Economic Review*, Vol. 88, No. 2, Papers and Proceedings of the Hundred and Tenth Annual Meetings of the American Economic Association, May 1998, pp. 179–184.

———, "Saver Behavior and 401(k) Retirement Wealth," *American Economic Review*, Vol. 90, No. 2, Papers and Proceedings of the Hundred Twelfth Annual Meetings of the American Economic Association, May 2000, pp. 297–302.

President's Commission on Military Compensation, *Report of the President's Commission on Military Compensation*, Washington, D.C.: U.S Government Printing Office, April 1978.

Rooney, Jo Ann, testimony before the House Armed Services Military Personnel Subcommittee, February 28, 2012.

Stern, Steven, "Simulation Based Estimation," *Journal of Economic Literature*, Vol. 35, 1997, pp. 2006–2039.

Sullivan, Paul, "A Dynamic Analysis of Educational, Occupational, and Inter-Firm Mobility Decisions," draft manuscript, University of Virginia, October 2003.

Train, Kenneth, *Discrete Choice Methods with Simulation*, Cambridge, Mass.: Cambridge University Press, 2003.

U.S. Department of Defense, *Sixth Quadrennial Review of Military Compensation Vol. 1B*, U.S. Government Printing Office, Washington, D.C., 1988.

———, *Seventh Quadrennial Review of Military Compensation*, U.S. Government Printing Office, Washington, D.C., 1992.

———, *Military Compensation Background Papers, Fifth Edition*, U.S. Government Printing Office, Washington, D.C., September 1996.

———, *10th Quadrennial Review of Military Compensation, Vol. 1: Cash Compensation*, Washington, D.C., U.S. Government Printing Office, February 2008.

———, *11th Quadrennial Review of Military Compensation*, Washington, D.C., U.S. Government Printing Office, 2012.

Warner, John T., "More Evidence on the Personal Discount Rates of Military Personnel," unpublished manuscript, Clemson University, 2005.

Warner, John T., and Saul Pleeter, "The Personal Discount Rate: Evidence from Military Downsizing Programs," *American Economic Review*, Vol. 91, No. 1, 2001, pp. 33–53.

Williamson, Stephanie, *A Description of U.S. Enlisted Personnel Promotion Systems*, Santa Monica, Calif.: RAND Corporation, MR-1067-OSD, 1999. As of September 28, 2012: http://www.rand.org/pubs/monograph_reports/MR1067.html

Wooldridge, Jeffrey A., "Simple Solutions to the Initial Conditions Problem for Dynamic, Nonlinear Panel Data Models with Unobserved Heterogeneity," *Journal of Applied Econometrics*, Vol. 20, No. 1, January 2005, pp. 39–54.